Luís Carlos Araújo Moraes

Environmental Licensing

Luís Carlos Araújo Moraes

Environmental Licensing

A conflicting vision in the preservation of natural resources

ScienciaScripts

Cover image: www.ingimage.com

This book is a translation from the original published under ISBN 978-3-330-75947-3.

Publisher:
Sciencia Scripts
is a trademark of
Dodo Books Indian Ocean Ltd. and OmniScriptum S.R.L publishing group

120 High Road, East Finchley, London, N2 9ED, United Kingdom
Str. Armeneasca 28/1, office 1, Chisinau MD-2012, Republic of Moldova, Europe
Printed at: see last page
ISBN: 978-620-8-30999-2

SUMMARY

To my wife for her patience and understanding of the hours I spent working on this research, giving up our leisure time.

ACKNOWLEDGMENTS

First of all, I thank God for giving me the opportunity to learn every day and thus the chance to improve and evolve in my life.

To my advisor, Dr. Fernanda Carla Wasner Vasconcelos, for her guidance, encouragement and patience throughout the development of this work.

To the technical team of AMAR's Environmental Licensing Department, especially analysts Joao Carlos Ferreira and Rosangela Azevedo Sardinha Duque, for their help and support in the research carried out.

To my colleagues in the master's program for their friendship throughout the course.

Finally, thank you to everyone who contributed in some way to my master's degree, even if they are not among the names mentioned here.

SUMMARY

The paper presents a brief theoretical discussion of the National Environmental Policy and its control mechanisms, with emphasis on Environmental Impact Assessment - EIA, Preliminary Environmental Impact Studies, Environmental Licensing, the subject of this study, and Ecological and Economic Zoning - EEZ. Next, the importance of biodiversity and natural resources was discussed, along with the tools for their conservation and preservation. It then emphasized the conflicting nature of legislation on land use and occupation. This study aims to analyze the environmental licensing of real estate projects as an instrument for preserving biodiversity and natural resources in the municipality of Resende (RJ) in the years 2009-2011. The methodology used was to identify and separate the administrative environmental licensing processes that were filed with the Resende Environment Agency during this period, and to select three real estate developments for a case study. These developments were chosen because they are located in fragile areas and were selected using the Resende-Geo GIS. The result points to the contradictory process of licensing by the municipality as a tool for preserving natural resources.

Key words: real estate developments, environmental licensing, natural resources.

1 INTRODUCTION

According to Primack & Rodrigues (2006), biodiversity or ecological diversity includes all interactions between biological components, formed by living beings, and abiotic components, such as air, water and climate.

According to Article 2 of the Convention on Biological Diversity - CBD, biological diversity means "the variability of living organisms from all sources, including, but not limited to, terrestrial, marine and other aquatic ecosystems and the ecological complexes of which they are part; including diversity within species, between species and of ecosystems" (MMA, 1992).

The greatest biological diversity on Earth is found in the world's tropical regions. The Atlantic Rainforest is the second largest tropical rainforest in Brazil, second only to the Amazon Rainforest. It used to extend from Rio Grande do Norte to Rio Grande do Sul, occupying around 15% of the national territory. Today, approximately 7% of its original native vegetation remains (FERRAZ, 2010).

The state of Rio de Janeiro alone preserves around 20.33% of the remaining Atlantic Forest biome in Brazil, covering an area of approximately 900,000ha (INSTITUTO BIOMAS, 2009). Due to its high biodiversity and endemism, this is a strategic area for conservation, not least because the majority of species, both fauna and flora, have not yet been identified or described.

Due to its importance, the Atlantic Forest biome is protected by Federal Law No. 11.428/06, which includes the preservation of biodiversity, the sustainable use of natural resources and the re-establishment, or even restoration, of degraded areas (BRASIL, 2006 b).

Another relevant factor for conserving the biodiversity of this biome is that studied by environmental economics, which is a branch of economics concerned with the relationship between animal and plant species and the products derived from them for food, clothing, housing and pharmacology, adding economic value to biodiversity.

For Primack & Rodrigues (2006), environmental economics, which integrates economics, environmental science, public policies and the values of biological diversity in its economic analysis, should give direct and indirect economic values to biodiversity and natural resources.

In this sense, water quality, soil protection, the interaction of wild species with commercial plantations, the practice of ecotourism, new medicines, biological control agents are some of the environmental services that are among the values mentioned above.

However, the physiognomy of cities has been undergoing changes due to rapid urban development. Urban vegetation is being eliminated on the banks of streams and rivers, along with hilltops, and areas with steeper slopes are being occupied, giving way to housing developments. These practices have contributed to increased environmental degradation.

Brazil's occupation process was characterized by a lack of planning and the consequent destruction of natural resources. The disorganized exploitation of Brazilian territory is one of the main causes of species extinction (IBAMA, 2007). Irresponsible consumption of water resources, silting up and eutrophication of rivers and lakes are examples of people not being aware of their role in the hydrological cycle. Clean, healthy water is essential not only for human beings, but also for life on Earth, as well as being a limiting resource for socio-economic development.

As such, real estate developments interfere with the landscape in a very significant way, and must therefore be subject to environmental licensing in order to guarantee the environmental conservation and preservation described in the law.

Article 225 of the Federal Constitution states that everyone has the right to a healthy environment and that, since it is a common good, it is everyone's responsibility to preserve it for current and future generations (BRASIL, 1988).

According to Brazilian legislation, the environment is classified as public property, so it must be protected for the good of the community, and it is up to the public authorities to legislate on its planning, use and occupation (MILARÉ, 2004).

To this end, legislation provides for a series of control instruments through which the possibility and regulation of any and all anthropogenic intervention on the environment can be checked. According to Martins (2007), the expansion of urban areas is one of the main causes of riparian forest degradation.

The implementation of allotments is an activity that modifies the environment and has the potential to cause significant environmental impacts, both in the natural environment and in the built, urban environment.

In this sense, environmental licensing is a legitimate instrument for managing the environment, since through it, the public administration exercises control over human activities that interfere with environmental conditions, in order to find a balance between social and economic development and environmental preservation.

There are various tools used to preserve and conserve biological diversity, such as the establishment of protected areas, known as Conservation Units, instituted by the National System of Conservation Units - SNUC, in Law No. 9.985/00 (BRASIL, 2000), through environmental licensing of activities with a high polluting potential and current legislation directly related to these processes, such as: Ecological Economic Zoning, the Forest Code, the Environmental Crimes Law, Resolutions of the

National Environment Council - CONAMA, among others.

The relevance of this study lies in the concern about the irregular use and occupation of land in the municipality of Resende, state of Rio de Janeiro, which has been degrading natural resources, as well as contributing to the reduction of biological diversity, given its importance for the evolution and maintenance of the systems necessary for life in the biosphere.

In this sense, the question is whether the environmental licensing process for real estate developments in practice presents conflicts between the relevant legislation and the protection of biodiversity and natural resources, and how this process has contributed to the conservation and preservation of biological diversity and natural resources.

2. GENERAL OBJECTIVE

- To analyze the environmental licensing process, as a tool for preserving biodiversity and natural resources, for real estate developments in the municipality of Resende, state of Rio de Janeiro, between January 2009 and July 2011.

2.1 SPECIFIC OBJECTIVES

- To check whether environmental licensing includes an environmental impact assessment aimed at preserving the Atlantic Forest biome in the municipality of Resende - RJ.
- Propose guidelines that promote the conservation of natural resources in the environmental licensing process for real estate developments.

3. THEORETICAL FRAMEWORK

3.1 NATIONAL ENVIRONMENTAL POLICY

The relationship between laws and ecosystems has consequences for biodiversity. In global terms, the 1972 UNESCO Convention considers natural assets to be the common good of humanity and, therefore, they must be preserved, as they are unique and irreplaceable for whatever people they belong to (UNESco, 1972).

With the enactment of Law No. 6.938/81 (BRAZIL, 1981), the country formally had a National Environmental Policy, a legal framework for all public environmental policies to be developed by the federal entities, with the integration and harmonization of these policies based on the objectives and guidelines established in this legal provision.

Law no. 6.938⁄81, which established the National Environmental Policy (PNMA), provides important definitions of the environment, degradation of environmental quality, pollution and environmental resources, and also instituted an important environmental protection mechanism called environmental impact assessment (EIA), These are effective and modern instruments in global environmental terms, especially with regard to the protection of ecosystems and the rational use of soil, subsoil, water and air (BRASIL, 1981).

The PNMA established the National Environmental System - SISNAMA, the National Environmental Council - CONAMA, the objectives and the instruments for controlling this policy. SISNAMA was established by Article 6 and its purpose is to establish a network of government agencies at the various levels of the federation, with the aim of ensuring mechanisms capable of efficiently implementing the PNMA. It is made up of a higher body - the Council of Governments; a consultative and deliberative body - CONAMA; a central body - the Ministry of the Environment; an executing body - IBAMA and various sectoral bodies, such as state, sectional and local bodies, such as the Municipal Environment Secretariats, each of which has its own competencies within this system (BRASIL, 1981), as shown in figure 1 of the organizational chart approved by Decree No. 6.101/07 (BRASIL, 2007).

FIGURE 1 - MMA'S ORGANIZATIONAL STRUCTURE

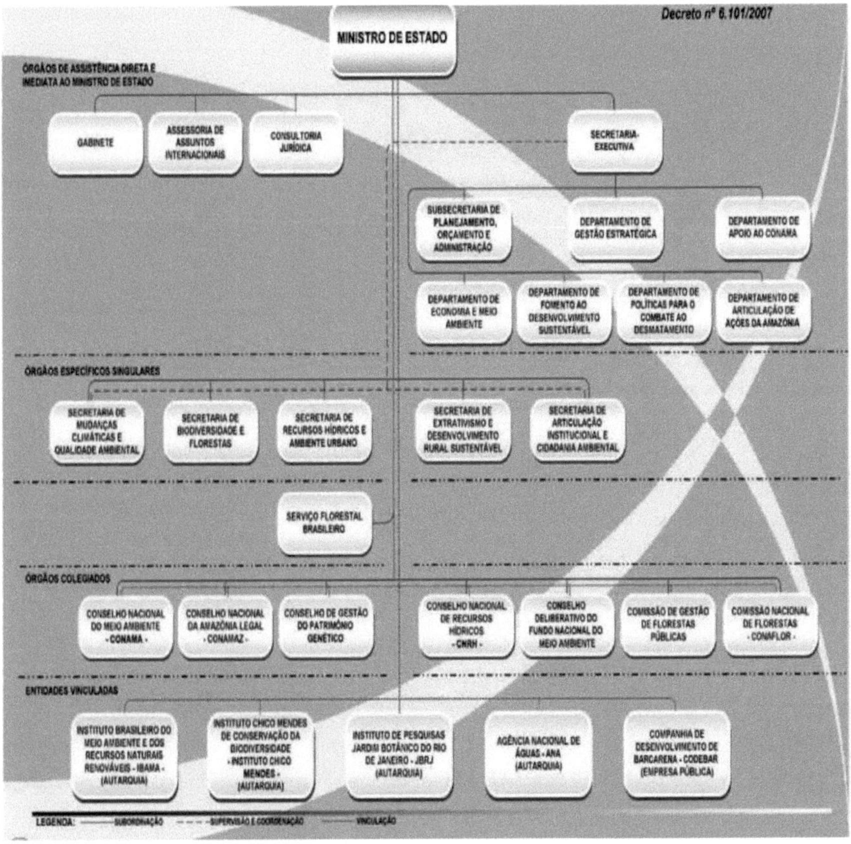

Source: Ministry of the Environment - MMA, 2007.

Art. 2^0 defines the objective as "the preservation, improvement and recovery of environmental quality conducive to life, with a view to ensuring conditions for socio-economic development in the country, the interests of national security and the protection of the dignity of human life" (BRASIL, 1981), where preserving means perpetuating, to leave environmental resources untouched, to improve is to make environmental quality continuously better through the rational management of these resources, and to recover is to seek the *status quo* in order to return degraded areas to the environmental characteristics they had before anthropic action (FARiAS, 2006).

Therefore, it is understood that the aforementioned law aims, among other things, to preserve and use natural resources in a rational manner, and that the instrument for achieving this goal is the licensing of activities that are effective or potentially polluting, as well as the assessment of impacts resulting from these activities.

As you can see, this objective is far-reaching, but this policy, in its Article 4, presents several other objectives, making it a set of legal, technical, scientific, political and economic instruments aimed at promoting sustainable development (BESSA, 2005).

In Article 8, the PNMA establishes CONAMA's legal powers, the purpose of which is to advise, study and propose to the higher body, government guidelines and policies for the environment and natural resources and also to decide, within the scope of its competence, on norms and standards compatible with the ecologically balanced environment and essential to a healthy quality of life (BRASIL, 1981).

According to Villares (2008), the importance of CONAMA lies in its composition, the involvement of the State, the productive sector and environmental organizations in establishing environmental quality standards, which are materialized through Resolutions, such as No. 237/97, which deals with environmental licensing (BRASIL, CONAMA, 1997).

Article 9 of the PNMA lists thirteen control instruments, some of which were created in 1981, but their wording has changed over the years, and others have been created in the light of environmental legislation case law. These include the assessment of environmental impacts, the licensing of effectively or potentially polluting activities and environmental zoning (BRASIL, 1981).

3.1.1 Control Instruments for Environmental Management

Among the control instruments mentioned in Article 9 of Law No. 6.938/81, emphasis will be placed on Environmental Impact Assessment - EIA, Prior Environmental Impact Studies, Environmental Licensing, which is the subject of this study, and Ecological and Economic Zoning - EEZ, where the former is necessary for the use of the latter instrument.

In fulfillment of its mandate, CONAMA established Resolution No. 001/86, which established the criteria and guidelines for the preparation of the EIA/RIMA, which subsidize the environmental licensing process for activities considered to have a high polluting potential and, in its article 1, defined environmental impact as

> any alteration of the physical, chemical and biological properties of the environment, caused by any form of matter or energy resulting from human activities that directly or indirectly affect it:
>
> I - the health, safety and well-being of the population;
>
> II - social and economic activities;
>
> III - the biota;
>
> IV - the aesthetic and health conditions of the environment;
>
> V - the quality of environmental resources (BRASIL, CONAMA, 1986).

Impacts can be qualified and, where possible, quantified as positive or negative for human conditions and the environment. Furthermore, in extreme conditions, they need to be compensated for.

According to Medeiros (1995), the EIA should initially be conceived as a preventive instrument of public policy and only becomes efficient when it is an aid to decision-making, being a tool for planning and designing projects so that sustainable development can be achieved.

According to Sousa (2006), EIA is a quantitative and qualitative analytical tool that aims to predict the impacts caused by anthropogenic action, as well as methods for mitigating negative impacts. Therefore, EIA is an extremely important tool for studying the use of natural resources, with a view to using them responsibly for future generations, thus entering the sustainability framework.

The EIA is associated with the environmental licensing process by describing environmental impacts such as ecosystem and landscape degradation, loss of vegetation cover, physical, chemical and biological alterations to the body receiving effluents, their magnitude, as well as the mitigating measures of the activities so that they can serve as a basis for granting the appropriate licenses.

In this sense, art. 10 of Law No. 6.938/81 prescribes that

> the construction, installation, expansion and operation of establishments and activities that use environmental resources, considered to be effectively or potentially polluting, as well as those capable, in any form, of causing environmental degradation, will depend on prior licensing by the competent state agency, part of the National Environmental System - SISNAMA, and by the Brazilian Institute for the Environment and Renewable Natural Resources - IBAMA, on a supplementary basis, without prejudice to other licenses required (BRASIL, 1981).

Land development for urban purposes, in any of its forms, is considered to be an undertaking potentially capable of causing environmental degradation and, as such, is subject to prior environmental licensing, according to Annex I of CONAMA Resolution No. 237-1997. Therefore, real estate developments or urban land subdivisions are activities considered harmful to the environment, since they alter the biological, aesthetic and sanitary properties of the environment. For this reason, environmental impact studies and their respective reports must be approved by the environmental agency before installation and operation activities can begin. These activities will be licensed in their various stages (BRASIL, 1997).

In some cases, such as enterprises or activities that do not have a great capacity to generate environmental impacts, the Environmental Control Report (RCA) is the document required for the licensing of the enterprise, in lieu of the EIA/Rima or during the Installation License. It is through the RCA that the entrepreneur identifies the actual or potential non-conformities resulting from the installation and operation of the enterprise for which the license is being requested.

The Environmental Control Plan (PCA) is the document through which the entrepreneur presents the plans and projects capable of preventing and/or controlling the environmental impacts resulting from the installation and operation of the real estate project for which the license is being requested, as well as correcting the non-conformities identified. In the state of São Paulo, the PCA is always required and is requested during the Installation License (SÂO PAULO, 2010).

In the state of Rio de Janeiro, state legislation limits itself to saying that urban development projects in areas over 50ha, or smaller when they are adjacent to Conservation Units, must present an EIA/RIMA.

As for the licensing of anthropic activities that use environmental resources, this was regulated by CONAMA Resolution No. 237/97. This is a procedure based on the Precautionary Principle, which determines that no intervention in the environment should take place before it is certain that it will not be adverse to the environment.

According to Gusmao (2004), the Precautionary Principle aims to prevent damage in relation to the known consequences of carrying out a given act, and the Precautionary Principle is used when the consequences of a given act are not known for certain, i.e. it is imperative when there is a lack of absolute scientific certainty. This lack of certainty cannot be dismissed as a reason for not adopting effective measures to prevent degradation.

The precautionary principle is directly linked to the quest to protect the environment, as well as the integrity of human life. This principle seeks to act in advance of environmental damage occurring. In this sense, Milaré (2004, p.144) informs us that "precaution is a noun from the verb precaver-se (from the Latin *prae* = before and *cavere* = take care), and suggests advance care, caution so that an attitude or action does not result in undesirable effects".

The purpose of environmental licensing is to minimize and compensate for the impact caused by activities considered to have a high polluting potential. It is, therefore, an instrument provided for by law and used by the public authorities for environmental control and is defined by CONAMA Resolution No. 237/97 as an administrative procedure by which the competent environmental body licenses the location, installation, expansion and operation of undertakings and activities that use environmental resources and are considered to be effectively or potentially polluting or those that, in any form, may cause environmental degradation, taking into account the legal and regulatory provisions and technical standards applicable to the case (BRASIL, 1997).

The environmental license is a binding, unilateral administrative act of the executive branch, with a defined period of validity, in which the environmental agency establishes rules, conditions, restrictions and environmental control measures to be followed by an enterprise in terms of its location, installation, expansion and operation.

Permits are the subjective right of the interested party and, as such, cannot be denied by the public administration, provided that the interested party complies with the legal requirements established by the environmental agency.

Environmental licensing is structured in three mandatory stages: Preliminary License (LP), Installation License (LI) and Operating License (LO) corresponding to the different phases of a project's implementation. The licenses are sequential, independent and linked to other federal, state or municipal licenses.

The LP is granted during the planning phase of the project, approving its location and establishing the basic requirements and conditions to be met in the next phases. The LI authorizes the installation of the enterprise or relevant activity in accordance with the specifications contained in the previously approved projects and the LO authorizes its operation.

The environmental license is temporary, because environmental issues are dynamic, so its renewal is not automatic. Generally speaking, the term of validity of the LP must be at least that established by the schedule for drawing up the plans, programs and projects relating to the enterprise or activity, and may not exceed five years. The validity period for the LI must be at least that established by the installation schedule for the undertaking or activity, and may not exceed six years. The validity period of the LO must take into account the environmental control plans and will be a minimum of four years and a maximum of ten years (BRASIL, 1997).

It can also be seen that a license is revoked when relevant information that was used to issue the license is omitted or falsely described. This happened at the beginning of 2010 in Cidreira, Rio Grande do Sul, where an electricity generation plant was installed on the dunes of the Salinas resort. The reason for the Curicaca Institute's request to revoke the license was the environmental importance of the area in which the project was licensed (CURICACA, 2010).

According to the Institute, this is a huge dune field, covering some 3,000 hectares, where the entire sequence of coastal environments - beach, frontal dunes, large dunes, marshes and lagoons - with associated animals and plants - has been preserved. One of them is the tuco-tuco (*Ctemomys brasilliensis*), an endemic and endangered species, which is why it has been classified as an area of extreme priority for the conservation of biodiversity by the

Ministry of the Environment and indicated for the creation of a Nature Conservation Unit in the Park category, by the State Department of Forests and Protected Areas of the State Secretariat for the Environment.

Most states of the federation follow CONAMA Resolution 237/97 in their licensing processes, with an increase in the state of Minas Gerais, which mentions it in the case of projects located in APAs, listed areas and areas of special tourist interest. In all states, there is also the need to present some kind of environmental study prior to licensing, such as: the Environmental Control Plan and Project - PCA; the Environmental Control Report - RCA; the Preliminary Environmental Report - RAP; the Environmental Diagnosis; the Management Plan; the Degraded Area Recovery Plan - PRAD or the Preliminary Risk Analysis (MILARÉ, 2004).

The Federal Constitution establishes that responsibilities for environmental issues should be shared between the federal entities. In this sense, the State Environmental Institute (INEA) in Rio de Janeiro has a policy of decentralizing environmental licensing, as it does in the states of Minas Gerais and São Paulo and, through an agreement, gives municipalities responsibility for licensing and inspecting activities with a local impact, according to Decree No. 42.050/09 (RIO DE JANEIRO, 2009).

It is worth noting that in the state of Rio Grande do Sul, Porto Alegre was the first city in Brazil to have a Municipal Environmental Department and that the state serves as a reference since it was also the first to decentralize the environmental licensing process in 1996. (AGNES et al; 2009).

However, the state of Rio de Janeiro was a pioneer in establishing environmental policies in the country, even before the publication of the National Environmental Policy (Law No. 6.938/1981), and its environmental licensing process was regulated by State Decree-Law No. 134/75 and State Decree No. 1.633/77, serving as a model for other states (RIO DE JANEIRO, 1975; 1977).

In an innovative way, State Decree-Law No. 134/75 proposes the obligation of prior authorization for the operation or functioning of its facilities or activities that are actually or potentially related to environmental pollution. The decree also established penalties in the event of pollution or non-compliance with the established rules.

This decree was partly regulated by State Decree No. 1.633/77, which established the Licensing System for Polluting Activities - SLAP (RIO DE JANEIRO, 1977), which already in 1977 provided for the issuing of preliminary, installation and operating licenses. More than a decade later, State Law No. 1.356/88 detailed the projects subject to the preparation of a Preliminary Environmental Impact Study (EIA) and its report (Rima) (RIO DE JANEIRO, 1988).

For real estate developments, there are basically the following stages in the procedure for obtaining licenses: The procedure must begin with a topographical survey of the area, which must be sent for an assessment of its urban viability and the establishment of land use guidelines, which have the function of planning the territory. After the Works Department has given its opinion on these issues, the environmental feasibility of the project must be assessed by the competent environmental agency, which, if it concludes positively, will grant the Prior License. Once the urban planning and environmental guidelines have been obtained, the urban planning and technical projects (water supply, electricity, sewage, etc.) must be drawn up and approved by the municipality. Once the projects have been accepted and the environmental license has been obtained, the environmental control measures and other conditions are defined. Finally, an urban planning license is required for the construction of the project.

According to the Federal Constitution, it is the responsibility of municipalities, through their Master Plans, to promote appropriate territorial planning, by planning and controlling the use, subdivision and occupation of urban land (BRASIL, 1988).

The Master Plan, which is mandatory for cities with more than 20,000 inhabitants, aims, among other things, to prevent pollution and environmental degradation (BRASIL, 2001a).

Thus, the three levels of government are empowered to license undertakings with environmental impacts, ratifying the proposal of the National Environmental System - SISNAMA, which encourages administrative decentralization in a cooperative and harmonious manner, and it is only up to the federal, state or municipal body to determine, according to the type and characteristics of the undertaking, who is responsible for this process.

Article 19, item I of Decree 99.274/90, which regulates the PNMA, requires compliance with municipal, state and federal land use regulations in order to obtain a Prior License (LP). In addition, the environmental licensing process must include a certificate from the City Council stating that the site and the type of development are in compliance with the legislation applicable to land use and occupation, in accordance with CONAMA Resolution No. 237/97 and the current IBAMA Normative Instruction No. 184/08. It can therefore be concluded that the Master Plan helps in the licensing process for real estate projects (BRASIL, 1990a).

Last but not least is the EEZ, which according to the Ministry of the Environment is an instrument used for planning and organizing the Brazilian territory so that economic, social and environmental relations are sustainable (BRASIL, 2002c).

In the state of Minas Gerais, the EEZ is used to draw up a diagnosis of the geo-biophysical, social, economic, legal and institutional environments, generating a map of environmental vulnerability and a map of social potential that will support priority actions for the protection and conservation of biodiversity and for development based on economic, social and ecological sustainability (MINAS GERAIS, 2002).

For Steinberger (1997), although maps are used, this is not their ultimate goal. Maps are used to help make decisions about which activities can be implemented in territories, thus minimizing their negative effects.

For Santos (2004), maps establish planning areas according to a systemic assessment of natural and socio-economic elements, the result of which is the development of rules for land use and occupation and the management of natural resources from a conservationist and economic and social development point of view.

Thus, in the state of Rio de Janeiro, the EEZ, regulated by State Law No. 5.067/07, took into account the state's ten hydrographic regions. The major river basins have their sources in the states of Sao Paulo, Minas Gerais and Espirito Santo, whose socio-economic activities affect the socio-environmental quality of the state of Rio de Janeiro (RIO DE JANEIRO, 2007).

The state of São Paulo was equally concerned about its coastline, considering in its EEZ the various ecosystems found in its coastal region due to the undesirable impacts of increased population density, tourist activity and real estate speculation in this area (SÂO PAULO, 2004).

Finally, it can be concluded that the control instruments presented here are of fundamental importance for the preservation of biodiversity, since in the near future biological diversity will play a key role as a source of raw material for the synthesis of new chemical products, especially for the pharmaceutical industry. Just remember that approximately 70% of medicines are made from plants (HARAGUCHI & CARVALHO, 2010).

3.2 LOSS OF BIODIVERSITY AND DEGRADATION OF NATURAL RESOURCES

The UN Conference on the Environment and Development, Rio-92, held in Rio de Janeiro in 1992, gave a global stamp to society's concern about environmental problems, reinforcing principles and rules for combating environmental degradation.

As a result of this Conference, five documents were approved, including Agenda 21, which is considered to be an action plan to be implemented by governments or any institution whose actions affect the environment (UN, UNCED, 1992).

Another document signed at the Conference and promulgated in Brazil through Federal Decree No. 2.519/98 was the Convention on Biological Diversity - CBD. The main concern of this Convention was the conservation of biodiversity, its sustainable use and the equitable sharing of the benefits generated by the use of genetic resources (BRASIL, 1998).

According to Santos & Góes (2006), considering the intrinsic value of biodiversity, reaffirming the sovereign right of states over their own resources and reaffirming their responsibilities for its conservation are the three basic principles of the CBD.

The Convention states that states have sovereign rights over their own biological resources and are responsible for the conservation of their biological diversity and the sustainable use of their biological resources (BRASIL, 1998). Following this orientation, each participating country is free to determine the legal, economic and political instruments that allow it to comply with the determinations described in art. 8 of the CBD. The guidance on these rights comes from the Federal Constitution, in its Article 23, on the common competence of the Union, the States, the Federal District and the Municipalities in sections III, VI and VII (BRAZIL, 1988).

According to the CBD (1992), biodiversity means the variability of living organisms from all sources, involving all the Earth's ecosystems and their interactions, and also includes diversity within species, between species and of ecosystems (BRASIL, 1992).

Biological diversity increases towards the Tropics, favored by the abiotic factors of temperature and humidity. According to Ricklefs (2010), organisms specialize in their *habitats*. The existence of an infinite number of *habitats* results in a large number of species living in them, which makes it impossible to quantify them.

There are believed to be between 10 and 50 million plant and animal species in the world. However, only 1.5 million have been classified by scientists to date. Around 20% of the world's known species are found in Brazil, which is considered a country of "megadiversity" (SÂO PAULO, 2010).

The richest biome in terms of biodiversity on the planet, the Atlantic Rainforest covers a total of 1,300,000km^2 , i.e. around 15% of the national territory, encompassing 17 Brazilian states and reaching areas such as Paraguay and Argentina. It is also worth noting that seven of Brazil's nine largest river basins support this biome (SOS MATA ATLANTICA, 2010).

With the creation of four new Conservation Units (UC) in June 2010, the state of Bahia now has 61% of its area protected. In total, these new conservation units will protect 65,070 hectares of Atlantic Rainforest. The newly decreed areas are part of the Central Atlantic Forest Biodiversity Corridor (CI - BRASIL, 2010).

Article 11 of the CBD recognizes that biodiversity has economic, social and environmental values, highlighting the need to use economic instruments to manage biodiversity conservation.

Sharing this vision, in September 2010, a business movement made up of companies such as Alcoa, Natura and Vale, as well as other institutions, launched a Business Charter for the Conservation and Sustainable Use of Biodiversity, which any company can adhere to, and sent a proposal to the Brazilian government suggesting at national level the establishment of mechanisms for the economic valuation of biodiversity and ecosystem services and their inclusion in Brazilian accounting legislation (INSTITUTO ETHOS, 2010).

Environmental economics - a branch of economics, currently confers direct economic values; those products that are harvested and used by people and indirect; those that do not involve the consumption or destruction of resources to the components of biological diversity (PRIMACK & RODRIGUES, 2006). In addition, Fenker (2007) adds that the economic value of a good is directly related to its scarcity, demand or need, so the more useful and scarce the good, the greater its economic value.

For the Brazilian Institute for the Environment and Renewable Natural Resources - IBAMA, the economic value of biodiversity consists of its use and non-use values. The former are made up of direct use values (recreation, leisure, harvesting of natural resources, hunting, fishing and education), indirect use values (watershed protection, habitat preservation for migratory species, climate stabilization, carbon sequestration) and option values (derived from the option to use the resource in the future); and non-use values include inheritance and existence values (IBAMA, 2010).

For Ricklefs (1996), the management of biotic resources must be done in a sustainable way, so it depends on ecological knowledge both to minimize environmental impacts and to indoctrinate economic, political and social thinking and practices.

Once biodiversity and natural resources are valued at their fair economic value, proportional to their usefulness, there will be an increase and incentive for conservation. No one will invest economic resources without a balanced return, because economic decisions are made taking into account the cost-benefit ratio (FENKER, 2007).

The economic aspect of water, for example, recognized in 1992 by the Dublin Declaration and adopted at the International Conference on Water and the Environment, in its fourth principle, states that water has economic value in all its competitive uses and should be recognized as an economic good (IETC, 2001), a fact ratified by the National Water Resources Policy (BRASIL, 1997).

With this in mind, the National Water Agency (ANA) instituted the Granting of the Right to Use Water Resources through Law No. 9.433/97, which ensures the quantitative and qualitative control of water use and the effective exercise of access rights. Charging for water use is a management instrument that came into operation in the federal basins in 2001 (BRASIL, 1997).

According to the Secretariat for Biodiversity and Forests of the Ministry of the Environment, anthropogenic actions such as the felling or burning of forests, the removal of mangroves and overfishing are factors that contribute to the destruction and loss of biodiversity (BRASIL, 2010).

In the 1990s, Sousa (1999) warned that various authors, such as Wilson, Ehrlick, Norgaard and Spears, corroborated the view that one of the main causes of the destruction of tropical forests is population growth and the consequent expansion of their activities necessary for survival.

Verdade, Dixo & Curcio (2010) add that deforestation for agriculture, air, water and soil pollution associated with disorderly exploitation of the environment, irrational use of natural resources and climate change have caused the destruction of *habitats* and are considered to be the main cause of species extinction.

Urban development is another factor that over the years has irreversibly compromised biodiversity and water resources. History shows that many cities have developed around watercourses, whether for navigation, for discharging effluents into their bodies or for human supply, such as the Danube River described by Ferraz (2010).

Increased industrialization, combined with urban settlements in marginal areas, are factors that affect water quality and cause the eutrophication of lakes and rivers, significantly compromising not only aquatic biodiversity, but also the multiple uses of water resources, surface water and groundwater, directly affecting the local economy.

In addition, the irregular occupation of hillsides, riverside areas and springs, stimulated by real estate speculation, is the major problem of water scarcity in the Southeast of the country, according to the National Water Agency - ANA (BRASIL, 2010).

Such is the importance of water resources that the United Nations Organization (UNO) created World Water Day (March 22), whose theme for 2011 was "Water for cities: responding to the urban challenge" (BRASIL, 2011).

In Brazil, the relationship with water resources happened in a different way from Europe, which used the river for navigation, enabling trade between countries. During the 18th century, squeezed between the mountains and the sea, the city of Rio de Janeiro grew and developed through successive embankments and drainage works, following the lines of the relief and with little or no planning (ABREU, 1992).

For historian Sérgio Buarque de Holanda, this sinuous urban layout is a reflection of the type of colonization undertaken by the Portuguese and their own worldview. For the author, the interest of the Portuguese in Brazil was to get rich quickly and with little effort. In his book Raizes do Brasil (Roots of Brazil), he quotes an excerpt from a letter written by Father Manuel de Nóbrega in 1552: "of those who came there, none has any love for this land [...] all want to do for themselves, even if it is at the cost of the land, because they expect to leave" (HOLANDA, 1995, p.151).

Continuing with disorderly growth at any cost, Minayo (1998) adds that around 1565, demographic growth took place in the form of subdivisions in sesmarias in the Guanabara Bay region, spreading out in various directions and eliminating various ecosystems.

According to Bueno (2004), irregular occupation is generally carried out in fragile areas such as riversides or hillsides. For the author, urbanization in their surroundings has always been an object of valorization - the construction of luxury condominiums on the banks of rivers - or of degradation of the urban landscape - the dumping and depositing of garbage, as occurs in the city of Recife (PE).

Another example of irregular occupation occurred in Belo Horizonte, where the urbanization process also took place in a disorganized manner, on hillsides, motivated by real estate speculation and without urban and social concern (SOUZA et al; 2010).

The process of urbanization, when carried out in a disorderly manner, causes a series of social and environmental problems, including unemployment, crime, slumification, exorbitant spending on public security and air and water pollution. According to the Brazilian Institute of Geography and Statistics - IBGE, there are 16,433 favelas in Brazil, most of them in Rio de Janeiro and Sao Paulo. A report by the Habitat Program, an organization linked to the United Nations (UN), also reveals that 52.3 million Brazilians (around 28% of the population) live in the country's registered favelas, reaching 55 million people by 2020 (VIVA FAVELA, 2010). Also according to IBGE (2008), approximately 16,267,197 people live in conditions of extreme poverty (53% of these in urban areas), 56% of households are not served by a general sewage system, 71.5% of municipalities do not have sewage treatment and 21.4% of households are not supplied with mains water.

According to Pinto (2003), "clandestine allotments", or "loteamentos clandestines", as they are known, are developments carried out on the fringes of the relevant legislation, in which streets are opened up and plots demarcated without any control from the public authorities.

Most of the time, these developments end in tragedy, such as the landslide that occurred on Morro do Bumba, in Niterói on April 7, 2010, where around 50 houses built on the site of a garbage dump collapsed, trapping hundreds of people, or the landslide in the Itajai Valley, in Santa Catarina, in November 2008, which, according to professors from the Universities of Santa Catarina, was caused by the disorderly occupation of PPAs, which was one of the reasons for the landslide (FRANK, 2008). In January 2011, the worst environmental disaster in the country occurred in the mountainous region of the state of Rio de Janeiro, with more than 800 deaths, due to the irregular occupation of fragile areas associated with the volume of rainfall.

Compans & Vinas (2005) warn that the population of this type of subdivision is exposed to environmental risks, in particular, flooding, landslides or mudslides, as well as contamination from the dumping of untreated sewage into water bodies.

According to Serra (2003), due to the speed of the urbanization process, the government is unable to control it, leading to a shortage of housing and infrastructure, resulting in an increasingly low quality of life.

According to Walcacer (1981), the lack of resources and unpreparedness to regulate the use and occupation of land means that town halls only become aware of an allotment when a few houses have already been built, and thus have no connection with the local reality, which only increases the socio-environmental impacts.

For the Rio Grande do Norte Institute for Economic Development and the Environment (IDEMA), the lack of an active environmental management body in most city halls, combined with the shortcomings in the state's technical-administrative structure to cope with the tasks, the absence of integrated planning and the existence of mechanisms to control land use and occupation, were the reasons that led the state government to implement the Estuaries EEZ in Rio Grande do Norte (RIO GRANDE DO NORTE, 2005).

In this context, rapid urbanization, combined with the lack of planning and the economic crisis, has led to total disorganization in land use and occupation, giving rise to neighbourhoods without any infrastructure, whose main characteristics are shantytowns, occupation on hillsides and river plains.

Some of these anthropogenic actions, or the sum of them, have contributed to the almost total devastation of the Atlantic Rainforest. It is estimated that around 93% of the original formation of this biome has already been devastated (SOS MATA ATLANTICA, 2010). The Cerrado, Brazil's second largest biome, occupies approximately 23.92% of the country's total area and has between 45 and 61% of its area preserved (IBGE, 2004).

In the Atlantic Forest biome, it is estimated that 55% of all tree species, 40% of non-tree species, 39% of mammals, including more than 15% of primates and 160 species of birds are also endemic to this region, i.e. they belong only to this geographical area (CI-BRASIL, 2010).

Among the species most vulnerable to extinction are those considered endemic and those in small populations. The smaller the population, the lower the genetic variability and the lower the variability, the greater the possibility of extinction.

In conclusion, biodiversity is the basis for agricultural, fishing and forestry activities and is strategic for the biotechnology industry. In other words, it has ecological, social, economic, scientific, educational, cultural, recreational and genetic value, as well as enhancing the local landscape. However, it is imperative that those responsible for environmental agencies and municipal inspection bodies make a greater commitment to complying with the Forest Code, obviously taking into account the urban area, which is already consolidated, because the reality of the facts is in dissonance with the public policies employed by the entities of the federation. This situation contributes to an even greater loss of biodiversity and the degradation of natural resources.

3.3 TOOLS FOR PRESERVING AND CONSERVING BIODIVERSITY AND NATURAL RESOURCES

Although widely used as synonyms, preservation and conservation are distinct concepts. The Brazilian Dictionary of Environmental Sciences defines preservation as "the action of protecting, against modification and any form of damage or modification, an ecosystem, geographic area, etc."

and conservation as "the rational use of any natural resource, while guaranteeing its renewal or self-sufficiency", and therefore differs from preservation in that it allows for the use and management of the area (LIMA-e-SILVA et al; 2002).

For Brazilian legislation, conservation means reconciling the rational use of natural resources with their protection, while preservation is the set of methodologies aimed at the long-term protection of species, *habitats* and ecosystems (BRAsIL, 2000).

Legislation also provides for various forms of preservation and conservation of biodiversity, found in the Forest Code, CONAMA Resolutions n°.237/90, n°.01/86, n°.302/02, n°.303/02 and the Environmental Crimes Law (MEDAUAR, 2011).

Thus, the sNUC (BRAsIL, 2000) and the CBD (UNCTAD, 1992) provide for the creation of Protected Areas or Conservation Units - CUs, as they are known - as the best way to protect biodiversity, since natural *habitats* are preserved and viable populations of species are maintained in their natural environment.

PAs represent the basic condition for the conservation and perpetuation of biodiversity (SÂO PAULO, 2010).

According to Faria (1997), Conservation Units are the most important form of biodiversity conservation, as they keep the Earth's evolutionary processes intact without human interference, thus leaving a legacy for future generations.

Bruner (2001) recognizes that the most effective proven strategy for promoting the maintenance of natural ecosystems is the implementation of efficient systems of protected areas.

Thus, according to Morin (2000), as the whole has characteristics that cannot be found in isolation, if these are isolated from each other, they will be inhibited by the restrictions coming from the whole. It is therefore necessary to conserve and preserve biodiversity in its natural environment.

Santos Filho (1995) also states that *in situ* conservation, that done *in* the natural environment, is recognized as the most effective method for maintaining the complex structure of ecosystems with their interactions between species, continuing the evolutionary processes that determine the variety of diversity.

In this sense, the environmental licensing of real estate developments must include in its planning the reduction of the isolation of vegetation fragments with programs for the recovery of native vegetation and stimulating conservation-friendly practices, as stated in Federal Law No. 11.428/2006 (BRASIL, 2006).

Regarding the importance of the areas surrounding Conservation Units, Primack & Rodrigues (2001) point out that if the areas surrounding Conservation Units are degraded, the biodiversity within them will decrease, especially if they are reduced in size, since many species move beyond the boundaries of the areas in which they normally live, in search of resources such as food and shelter.

The State of São Paulo, concerned about the growth of the human population, the occupation of previously forested areas, the destruction and degradation of *habitats*, the burning of oil and oil products for energy and transportation, and their effects on biodiversity, launched the Caderno de Educaçao Ambiental sobre a Biodiversidade (Environmental Education Notebook on Biodiversity) in 2010, a series of notebooks that still has three more titles published: Groundwater in the State of São Paulo, Ecocidadao and Nature Conservation Units (SÂO PAULO, SMA, 2010).

This initiative is undoubtedly due to the fact that the state of São Paulo has large remnants of the Atlantic Forest and intact areas of Cerrado, two biodiversity *hotspots* that are under threat (SÂO PAULO, SMA, 2010). A *hotspot* is an area with at least 1,500 endemic plant species and which has lost more than three quarters of its original vegetation, constituting a priority area for conservation (CI-BRASIL, 2005).

The conservation of natural ecosystems is one of the objectives of the SNUC, in Article 4, along with the maintenance of biodiversity and genetic resources (BRASIL, 2000). It is understood that only in this way will species be able to continue their evolutionary process, since the interactions between them and their habitat trigger the natural techniques of evolution.

Thus, the most feasible way to achieve these objectives, according to Santos Filho (1995), Morin (2000), Bruner (2001), and Primack & Rodrigues (2006), is to preserve communities in their natural environment, i.e. by creating legally protected areas (although creation alone does not guarantee protection of biodiversity), as provided for in Federal Law No. 9.985/2000, which established the National System of Conservation Units (SNUC) and defines a Conservation Unit as follows

> territorial space and its environmental resources, including jurisdictional waters, with relevant natural characteristics, legally instituted by the Public Power, with conservation objectives and defined limits, under a special administration regime, to which adequate protection guarantees apply (BRASIL, 2000).

In this way, one of the commitments contained in the CBD is assumed, which stipulates in Article 8 that member countries must establish a system of protected areas or areas where special measures need to be taken to conserve biological diversity.

The SNUC defines and regulates the categories of UCs at the federal, state and municipal levels, separating them into two groups: areas of integral protection, with the conservation of biodiversity as the main objective, and areas of sustainable use, which allow various forms of use of natural resources, with the protection of biodiversity as a secondary objective (BRASIL, 2000).

The following units are classified as sustainable use: Environmental Protection Area, Area of Relevant Ecological Interest, National Forest, Extractive Reserve, Fauna Reserve, Sustainable Development Reserve and Private Natural Heritage Reserve.

Integral protection units include Ecological Stations, Biological Reserves, National Parks, Natural Monuments and Wildlife Refuges. All of them have specific management plans, i.e. any procedure aimed at ensuring the conservation of biological diversity and ecosystems. The management plan must cover the area of the Conservation Unit, its buffer zone and ecological corridors, including measures to promote its integration into the economic and social life of neighboring communities (BRASIL, 2000).

A National Park - PARNA, for example, has the basic objective of

> the preservation of natural ecosystems of great ecological importance and scenic beauty, enabling scientific research and the development of environmental education and interpretation activities, recreation in contact with nature and ecological tourism (BRASIL, 2000).

Obstacles to the preservation of biodiversity have been identified in some PAs, such as the Itatiaia National Park in the state of Rio de Janeiro, which, although it falls into the category of full protection, has had hoteliers and residents inside since it was created. In this sense, Arruda (1997) points out that in 39% of the UCs in the states of Rio de Janeiro, Espirito Santo and Paranà there are residents. In Rio de Janeiro, 73% of them have residents in their internal area, while in Paranà, this occupation occurs in 20% of the UCs and in Espirito Santo, 25% of these UCs are used for housing purposes. In turn, the bi-state PAs have 39% of residents inside them.

In turn, the basic objective of an Environmental Protection Area (APA) is to ensure the sustainable use of its natural resources by regulating the process of land occupation, since a certain degree of occupation is permitted in its areas (BRASIL, 2000).

Without including private and municipal reserves, 52% of the Brazilian system of Protected Areas is federal, and 48% is under state jurisdiction. In Brazil, there are 310 federal Conservation Units and 367 state ones, with State Parks appearing mostly in the South and Southeast regions (RYLANDS & BRANDON, 2005).

However, data from the Ministry of the Environment shows that among the states that form the Atlantic Forest, Sao Paulo has invested the most in integral protection PAs - 23 State Parks, followed by Paranà with 19 and Rio de Janeiro with 8. The state of Bahia has invested the most in sustainable use PAs - 28, followed by Paranà with 9. Sao Paulo and Santa Catarina have not created any PAs and Espirito Santo has 5 PAs and 5 State Parks (BRASIL, MMA, 2010).

Another important tool for the conservation and preservation of biological diversity and natural resources in Brazil began in 1965 with Federal Law No. 4.771, known as the Forest Code, which defined Permanent Preservation Areas (APPs) as being

an area protected under the terms of articles 2 and 3 of this Law, whether or not covered by native vegetation, with the environmental function of preserving water resources, the landscape, geological stability, biodiversity, the gene flow of fauna and flora, protecting the soil and ensuring the well-being of human populations (BRASIL, 1965).

APPs include springs, hilltops and slopes, marshy areas and the banks of water bodies, all of which are considered to be fragile areas. These areas are considered to be of relevant environmental value due to the functions they perform and are therefore subject to a legal regime that must take into account the public interest.

The indiscriminate and uncritical occupation of the most fragile areas causes negative impacts on the environment, such as erosion, flooding, silting of watercourses and the elimination of springs, affecting the quantity and quality of water resources.

However, the occupation of PPAs is not exclusive to the poorest classes. Throughout the country, irregular building practices are being carried out by those with greater economic power in areas considered to be environmentally vulnerable, not only to meet their housing needs, but also to meet the demands of tourism.

In this context, the impact of the valorization of space by tourist activity has social, economic and environmental consequences. This is the case, for example, with the sealing of river banks in Bonito (MS) to facilitate tourist access and the installation of artificial lighting in caves to make them easier to see, the mega *resorts* installed on the coast of Bahia and the hotels installed on the coast of São Paulo (CRUZ, 2001).

The Costa do Sauipe tourist hotel complex was built within an Environmental Protection Area and surrounded by traditional populations. It is considered to be the largest tourist complex in South America, occupying an area of 1,755 hectares of the Sauipe farm, of which six kilometers are along the coast. The Sauipe farm was characterized by its abundance of water resources as well as having representative areas of the Atlantic Rainforest (RODRIGUES JUNIOR, 2006).

On the coast of São Paulo, a series of luxury hotels, horizontal condominiums and subdivisions for second homes have been built on remnants of the Atlantic Forest on the southern coast of the state. Similarly, in the state of Paranà, urban occupation continues from Paranaguà in the direction of the municipalities of Pontal do Paranà, Matinhos and Guaratuba, with tourism and real estate promotion being the main drivers, especially the establishment of marinas, horizontal condominiums in areas of mangroves and sandbanks, on the banks of the Boguaçu Mirim and Itiberê rivers (BRASIL, 2008).

In the urban context, APPs play an important role in improving the quality of life of residents, since they influence the local microclimate, reduce noise, dust and capture carbon dioxide, contribute to the hydrological cycle and improve the aesthetics of the landscape.

These functions prevent siltation of water bodies, thus avoiding flooding. In addition, vegetation cover reduces the amount of suspended materials that carry nutrients and toxic substances that reach the water, affecting its quality and affecting both humans and the aquatic ecosystem.

Martins (2007) corroborates this assertion, citing that riparian forests act as filters that retain the remains of pesticides, pollutants and sediments that would directly affect water quality. In addition, they connect forest fragments, forming ecological corridors, thus facilitating the movement of fauna, which contributes to gene flow between species.

For Magalhaes & Ferreira (2000), the role of the APPs within a sub-basin is evident and fundamental, as they preserve the various ecosystems with their biodiversity, responsible for the regular and permanent supply of drinking water from the springs as well as for protecting the natural resources necessary for life on Earth.

Thus, it can be concluded that even though PAs are a tool for conserving and preserving biodiversity, there is a conflict between their categories, serving both preservationist and socio-environmentalist objectives. As for the use of APPs, the conflict is even greater, and it is necessary to find a way of preserving an area, integrating the different categories of UCs and their respective buffer zones and ecological corridors together with the APPs, thus aiming to preserve biodiversity.

3.4 Conflict of laws

Despite the socio-environmental advantages of maintaining APPs, the Forest Code has some conflicts. Initially, it was used only in rural areas and began to be used in urban areas after Provisional Measure No. 2.166-67/01 (BRAZIL, 2001), a fact that is generating a great deal of controversy in current environmental law, given its legal provisions: Federal Law No. 6.766/79 (BRAZIL, 1979), which provides for the subdivision of urban land, as well as CoNAMA Resolution No. 369/06 (BRAZIL, 2006).

The Law on the Parceling of Urban Land, in its Chapter II, art. 4, III reads as follows: "along running and dormant waters and the public right-of-way of highways and railroads, it shall be compulsory to reserve a non-buildable strip of 15 (fifteen) meters on each side, unless otherwise required by specific legislation" (BRASIL, 1979). The Forest Code states that this preservation area should be 30 meters for watercourses less than 10 meters wide (BRASIL, 1965).

According to Servilha et al. (2005), the distances mentioned in Article 2 of the Forest Code for APPs have no scientific basis and exclude the existence of man and his influence on the dynamics of the landscape in which they are located.

In line with this information, the state of Rio de Janeiro signed Decree Law No. 42.356/10, which provides for a reduction in the measures adopted in the Forest Code in consolidated urban areas, leaving the delimitation of the marginal protection strip to the discretion of INEA, in line with federal legislation (RIO DE JANEIRO, 2010).

According to Lima (1996), there is still no definitive method for establishing the minimum width of riparian forest that will allow satisfactory protection of the watercourse.

In the legal sphere, there are various controversies regarding APPs, since Article 30 of the Federal Constitution states that municipalities are responsible for legislating on matters of local interest and supplementing federal and state legislation, where applicable (BRASIL, 1988).

It is worth noting that the National Congress is currently reformulating the Forest Code. The new bill tends to change conflicting points such as the distance between streams and springs in rural and urban areas, the size and form of legalization of legal reserves and the regularization of rural properties, with or without amnesty for those who caused deforestation for construction.

Currently, each state has a different understanding of the issue. In this sense, Porto Alegre's Master Plan, in 1999, in paragraph 3 of art. 88, allows only areas that maintain their original attributes to be considered Permanent Preservation Areas (PORTO ALEGRE, 1999).

In Rio de Janeiro, State Decree No. 42.356/10 in its Article 4 states that the minimum limits set abstractly by Article 2, "a", of the Forest Code (Federal Law No. 4.771/65 and its amendments) may be reduced in each specific case (RIO DE JANEIRO, 2010).

Article 2 of the Forest Code highlights the relationship between APPs and their use, within the various categories in urban areas. Thus, non-building areas are those along rivers or any watercourse from its highest level in a marginal strip whose minimum width varies from 30m to 500m depending on the width of the watercourse, areas around lakes, springs, hilltops and slopes with a gradient greater than 45° (BRASIL, 1965).

CONAMA Resolution No. 303/02, which regulates Art. 2 of Law No. 4.771/65, corroborates its measures with the Forest Code with regard to distancing from water bodies, as well as complementing it in Art. 3, providing measures for other APPs such as, for example, distancing in vereda and in a marginal strip, in horizontal projection, with a minimum width of 50m, from the limit of the marshy and waterlogged space (BRASIL, 2002b).

However, in 2006, CON AM A Resolution No. 369, considering the right to property as well as its socio-environmental function, in addition to the legal duty to recover APPs that had been irregularly suppressed or occupied by the owner, regulated the exceptional cases of public utility, social interest or low environmental impact that allowed the intervention or suppression of vegetation in APPs

(BRASIL, 2006). Article 9 of this Resolution sets out the requirements and conditions for intervening in or suppressing vegetation in APPs in the process of land regularization in urban areas, as defined below:

I - predominantly residential low-income occupations;

II - occupations located in an urban area declared a Special Zone of Social Interest-

ZEIS in the Master Plan or other municipal legislation;

III - occupation in an urban area that meets the following criteria:

a) have at least three of the following items of urban infrastructure in place: road network, rainwater collection, sewage disposal, solid waste collection, *water* supply network, energy distribution network;

b) have a population density of more than fifty inhabitants per hectare (BRASIL, CONAMA, 2006).

With regard to the impacts of real estate developments, Costa & Peixoto (2007), citing the case of *Alphaville*, in the Metropolitan Region of Belo Horizonte, question the treatment of sewage, the collection and final destination of waste, the effect on the flow of the lagoon with the increase in soil sealing, the impact on local fauna and others.

According to Jurisway Sistema Educacional (2009), allotment is a form of parceling out urban land, dividing it into smaller portions, with the aim of selling the parts to the public in successive and periodic installments. This is governed by Law No. 6.766/79, the Urban Land Allotment Law, which defines allotment as "the subdivision of a plot of land into lots intended for building, with the opening of new circulation routes, public roads or the extension, modification or widening of existing roads" (BRASIL, 1979).

Article 6 of CONAMA Resolution No. 01/86 stipulates, in one of the stages, that the environmental impact study should consider, among its technical activities, a diagnosis of the physical environment (water, air, climate, topography, among others), biological environment (fauna, flora and their ecological relationships) and socio-economic environment (land use and occupation) (BRASIL, CONAMA, 1986).

Finally, it can be concluded that despite the legal differences governing the matter, there is a consensus on the application of the Forest Code, including in urban areas. An example of this is the recommendation made by the Federal Public Prosecutor's Office to the Municipality of Resende (RJ) to use Law No. 4.771/65, since the municipality was using State Decree No. 42.356/10 to approve environmental licensing for real estate developments, to the detriment of the Forest Code.

4. METHODOLOGY

The region inventoried refers to the municipality of Resende (RJ), from January 2009 to July 2011, and the following procedure was used:

Firstly, the environmental licensing processes for real estate developments were identified and analyzed at the Environment Agency of the Municipality of Resende - AMAR, between January 2009 and July 2011. During this period, 350 cases of Prior Consultation on Environmental Impediments to real estate development were filed for analysis. Based on the agreement signed between the municipality and INEA, which determines which activities the municipality of Resende can license, the processes considered to have minimal environmental impact and small size and those with medium environmental impact and minimal, small and medium size were separated for analysis, according to Tables 1, 2 and 3, i.e. the municipality can license developments up to 500,000m^2 whose sum of the conditioning factors does not exceed the value of 35 in the sum of the parcels (Table 3).

Table 1 - Development Area

Size	Intervention Area (m^2)
Minimum	up to 2,000
Small	over 2,000 up to 20,000
Medium	over 20,000 up to 100,000
Large	over 100,000 to 500,000

Source: INEA, 2010.

Table 2 - Conditioning Factors

Weight	Conditioning Factor	Situation	Value
10	Located in or around a fragile area (Annex 1)	no	0
		yes	1
10	Provides for cuts and embankments	no	0
		yes	1
10	Provides for alterations to bodies of water or modifies natural drainage	no	0
		yes	1
8	Provides for the removal of vegetation	no	0
		yes	1

Source: INEA, 2010.

Table 3 - Determination of Pollution Potential

Polluting Potential	Sum of weight x value parcels
Bass	0 a 18
Medium	19 a 35

Source: INEA, 2010.

Administrative processes considered to be of minimal environmental impact and small in size go through a simplified licensing procedure in which the analyst makes a field visit to check whether or not the land has any legal environmental impediments to building, such as: proximity to streams, springs, marshy areas, hilltops and slopes in accordance with Article 2 of the Forest Code (BRASIL, 1965). In this case, the land may be totally or partially in APP or totally free to be built on. An example of this is Development A, the subject of this study, where there are plots totally, partially or free of environmental impediments.

The processes for projects considered to have a medium environmental impact and those of minimal, small and medium size, as determined by Tables 1, 2 and 3, are subject to a complete licensing procedure, in order to obtain the LP, LI and LO licenses.

After the analysis and issuance of the technical opinion, the environmental analysts file the documents electronically, separated by neighborhood, to facilitate later consultations. The same happens with administrative processes that have undergone full environmental licensing analysis.

In Resende-Geo, those processes that met the following criteria were identified: being located on a hilltop near a water body, a spring and a marshy area, areas considered to be permanent preservation. Thus, in accordance with these criteria and in order to emphasize the object of study, three developments considered to be the most critical in terms of the environmental aspects mentioned above were chosen for this case study, here entitled, developments A, B and C.

Resende-Geo is an environmental impact assessment tool, a Georeferenced Information System - GIS, developed for territorial management that allows working with alphanumeric spatial data and provides functional tools for analyzing, consulting, editing and restructuring data, generating reports and associating external data.

This GIS is part of an agreement called Programa de Gestao Ambiental e Fisico Territorial para Municipios (Environmental and Physical Land Management Program for Municipalities) signed between the Nùcleo de Assistència Tècnica aos Municipios - NAT (Technical Assistance to Municipalities) of the Pontifical Catholic University - PUC Rio, Petróleo Brasileiro S/A - Petrobràs and the Resende City Hall. This system has aerial images from 2005 to 2010 and is based on the mapping of the urban area covering a total area of $85km^2$ The aerophotogrammetric survey is of high spatial resolution (pixel= 25x25cm).

According to Uezu (2006), the importance of GIS in conservation comes from the need to understand how the location in space of the different elements that make up the environment and their interactions affect biodiversity.

Finally, to conclude the analysis, a comparison was made of the three allotments determined above, between 2009 and 2011, using images from Resende-Geo and updated digital photos from May 2011, taken *on site in* the study areas, using a digital camera, to check whether the developer had complied with the requirements of art. 2 of the Forest Code, respecting biodiversity and natural resources, as well as the CBD.This made it possible to check whether the licensing process was taking into account the preservation of biodiversity and natural resources as provided for in environmental legislation.

5. RESULTS AND DISCUSSION

5.1 STUDY AREA

5.1.2 Location and Characterization of the Study Area

Resende is a Brazilian municipality located in the south of the state of Rio de Janeiro, in the region of Vale do Mèdio Paraiba, Sul Fluminense, at an altitude of 395m, with a population of approximately 119,769 inhabitants (IBGE, 2010), 94% of whom are urban. It has a Tropical High Altitude climate with an average temperature of 25°C, and an average annual rainfall of 1,500mm. The highest rainfall is observed between December and March. The municipality of Resende (Figure 2) has a large territorial extension, 1,095.254km^2 (IBGE, 2010), the largest on the Rio-Sao Paulo axis, among the municipalities crossed by the highway linking the two metropolises.

FIGURE 2 - LOCATION OF THE MUNICIPALITY OF RESENDE, RJ

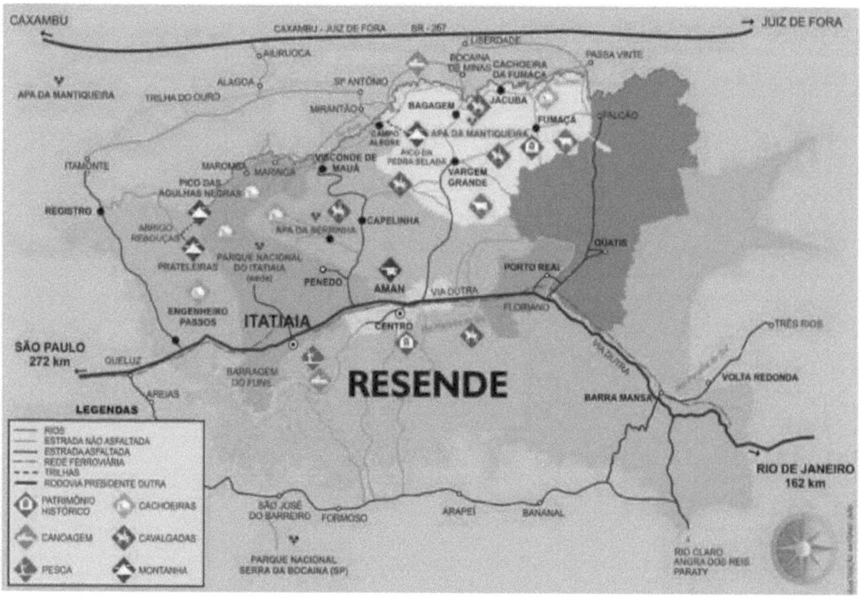

Source: Resende City Hall, 2010.

Its relief is typical of a valley, with the municipality located on a large plateau on the banks of the Paraiba do Sul River. As you move away from this area, you find a plateau with slight flattened hills and, further away, the Itatiaia Massif, which comprises an escarpment of the Serra da Mantiqueira, with the Pico das Agulhas Negras (2,787m) in the background, as shown in figure 3. At the other end of the municipality, near the São Paulo border, you find the beginning of the Serra do Mar formations, with the presence of elevations that generally exceed 600m in altitude.

FIGURE 3 - PARAIBA DO SUL RIVER AND ITATIAIA PARNA IN THE BACKGROUND

Source: Luis Araùjo, 2011.

The Rio de Janeiro Botanical Garden Research Institute (2010) lists 35 endemic species, 90 vulnerable species and 4 threatened species between the municipalities of Itatiaia and Resende, both in the state of Rio de Janeiro.

As such, SOS Mata Atlàntica (2008) identifies Resende as an extremely high priority area for the conservation of the Atlantic Forest biome, since there is only 22% of remaining original cover, as shown in figure 4.

FIGURE 4- REMAINING VEGETATION COVER IN THE MUNICIPALITY OF RESENDE (RJ)

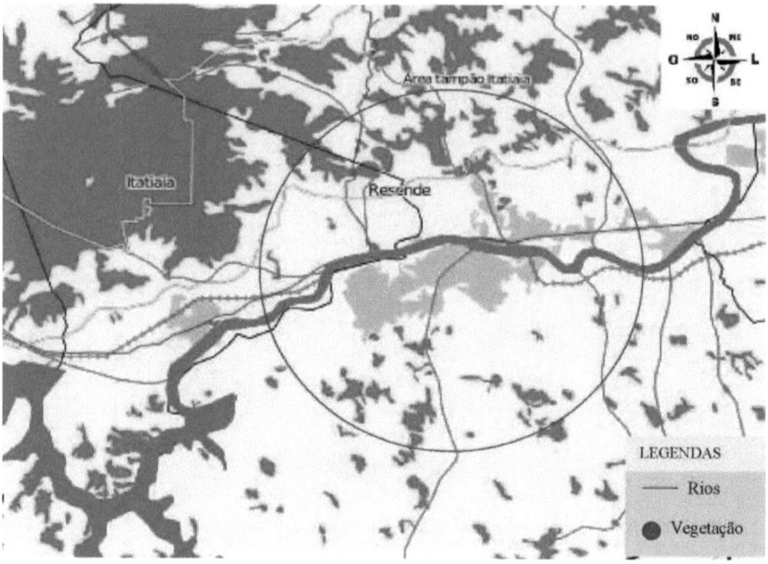

Source: SOS MATA ATLANTICA, 2008.

In this sense, a connection with the Itatiaia PARNA area and the Serra da Bocaina PARNA (sP) is important, as one of the greatest risks to forest remnants is the lack of connection between one fragment and another, reducing the so-called "edge effect"[1] .

According to Pinheiro & Viana (1998), when fragments are interconnected through biodiversity corridors, the flow of animals and seeds increases and, therefore, the colonization of degraded areas by the species of plants and animals present in the forest fragments.

This "edge effect" tends to negatively affect the biota, quantitatively and qualitatively reducing the biodiversity present in the CUs, posing risks to the management of these areas. Figure 5 shows irreplaceable areas for protecting Atlantic Forest species with biodiversity corridors.

FIGURE 5 - "IRREPLACEABLE" AREAS FOR SPECIES PROTECTION IN THE ATLANTIC FOREST

1 The term "edge effect" refers to changes in physical, chemical and biological parameters caused by anthropogenic or non-anthropogenic actions in the area where the vegetation fragment comes into contact with the surrounding matrix.

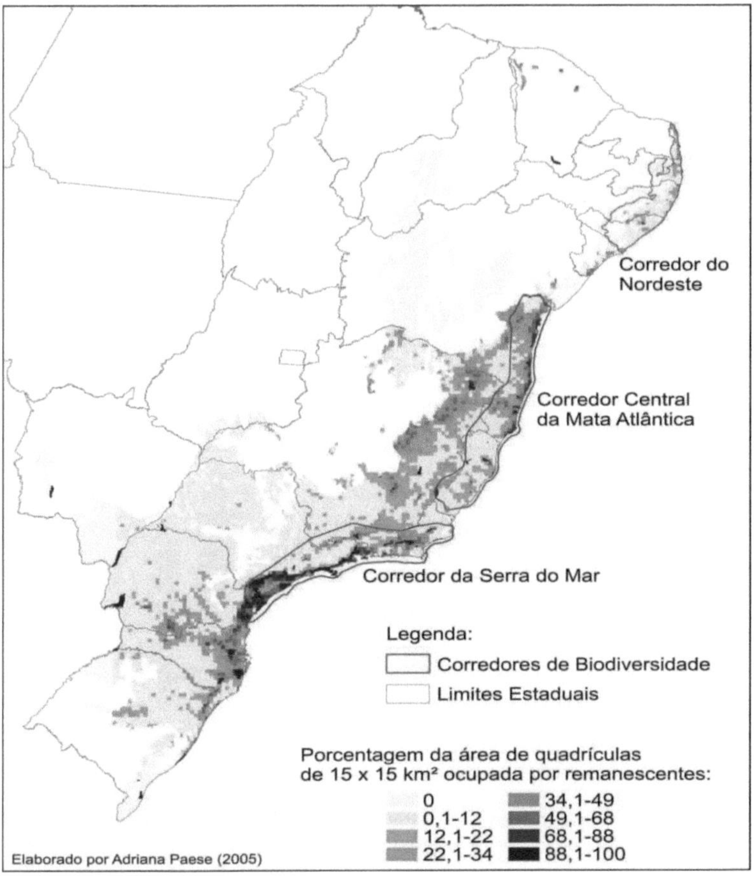

Source: Pinto, 2005

Figure 6 shows the distribution of Atlantic Forest remnants in their coverage area with their biodiversity corridor boundaries.

FIGURE 6 - DISTRIBUTION OF FOREST REMNANTS IN THE ATLANTIC FOREST AND BOUNDARIES OF BIODIVERSITY CORRIDORS

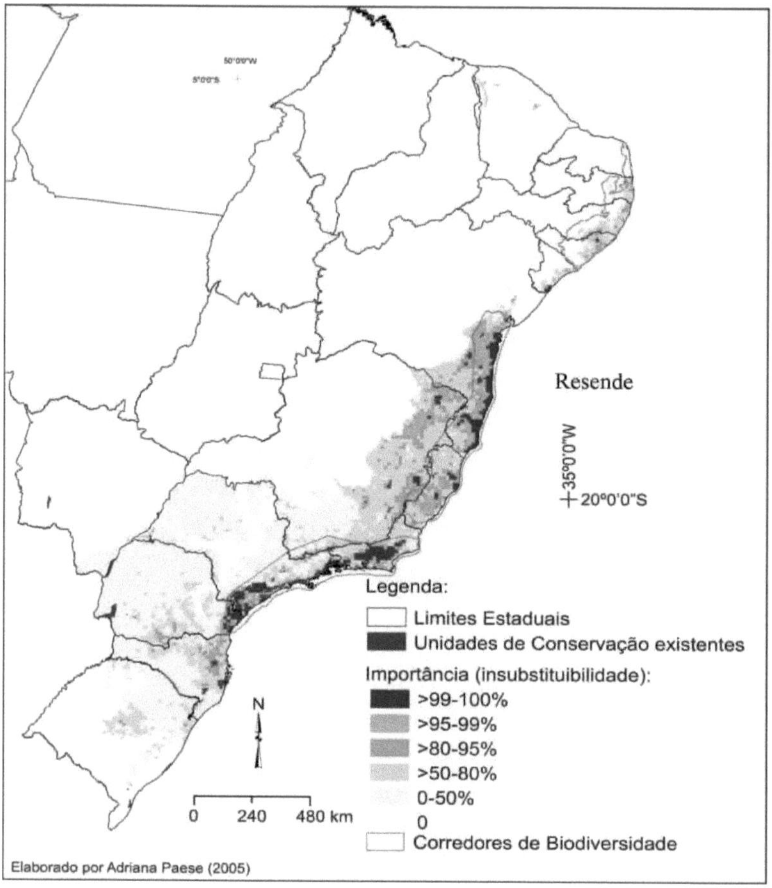

Source: Pinto, 2005

The rivers and lakes of the Atlantic Forest are still home to rich aquatic ecosystems, many of which are threatened by the removal of their riparian forests, giving way mainly to large real estate developments, without due environmental care. This tangled network of basins is made up of rivers of national and regional importance, such as the São Francisco, Paranà, Tietê and Paraiba do Sul.

The main tributaries of the Paraiba do Sul River within the municipality are the Alambari, Sesmaria and Lavapés rivers and the Preto stream. Of these, the Sesmaria River and the Paraiba do Sul River are under federal jurisdiction. Figure 7 shows the number of springs and water bodies in the municipality of Resende (RJ) and the location of developments A, B and C.

FIGURE 7 - AREA VIEW OF RESENDE WITH ITS WATERCOURSES

Source: Resende City Hall, 2009a.

5.2 TECHNICAL ASPECTS

Of the 350 cases of Preliminary Consultation of Environmental Impediment for real estate developments, 4% required a more specific environmental licensing process because they are medium and large environmental impact activities and the majority of requests underwent a simplified licensing process, since they have a small environmental impact.

FIGURE 8 - PERCENTAGE BY TYPE OF LICENSE

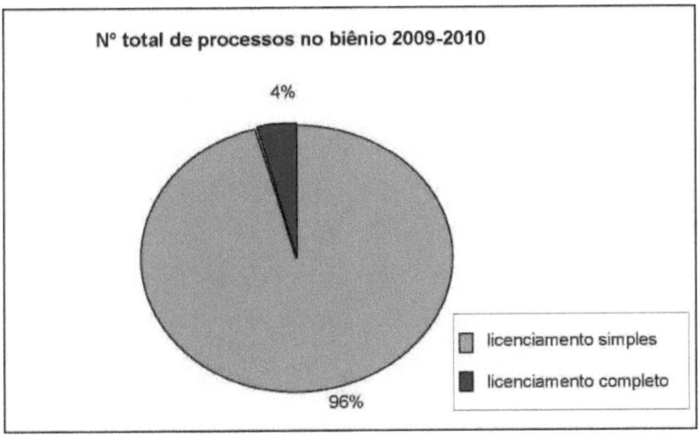

Source: AMAR, 2011.

Project A, as described above, underwent a simplified licensing process, in which all that is observed is whether or not the land is in an APP, in accordance with Article 2 of the Forest Code (BRASIL, 1965) and Article 3 of CONAMA Resolution No. 303/02 (BRASIL, 2002b).

For projects with a high environmental impact, projects B and C, it is necessary to comply with certain legal requirements, such as rainwater drainage plans, sanitary sewage, landscaping and demarcation of the marginal protection strip, among others, in accordance with the physical characteristics of the land, for subsequent release of the licenses: LP, LI and LO, due to their degree of impact, as set out in tables 1, 2 and 3 and in Annex 1.

The criteria used to delimit the APP in the case of streams, springs and hilltops were obtained from Article 2 of Federal Law No. 4771/65 (BRASIL, 1965), which determines the size to be respected in these cases, as well as Article 3 of CONAMA Resolution No. 303/02, which determines the distance from marshy areas (BRASIL, 2002b).

5.3 LEGAL ASPECTS

Firstly, it should be noted that the Municipal Environmental Agency (AMAR) has an agreement with the State Agency (INEA) to carry out the licensing process for real estate projects with a local impact, of small and medium to medium polluting potential (Annex 1) and, as of July 2011, the definition of who will be responsible for the licensing process for impact-generating activities, AMAR or INEA, is now done electronically through INEA's licensing portal via the link http://200.20.53.7/Ineaportal/Enquadramento/Passo1a.aspx.

In order to proceed with the licensing of any real estate development, it is necessary to present a declaration from the Resende City Hall (RJ) proving that the location of the development complies with the municipality's land use and occupation legislation, i.e. the licensed subdivision must comply with the municipality's regulations, such as the Master Plan, land use legislation and other rules.

However, it can be seen that the practice is not in line with current municipal regulations, as can be seen in Figures 9 and 10, highlighting the conflict that exists not only between the Forest Code, as the building is located in an APP (less than 30m from the riverbed) and CONAMA Resolution no.369/02, which provides for exceptional cases of public utility, social interest or low environmental impact, which allow the intervention or suppression of vegetation in APP, since the development is private, but also the conflict of competence in the licensing process between the State and the Municipality, since the State (INEA) was responsible for the licensing process for this development.

FIGURE 9 - CLINIC BUILDING AREA IN APP

Source: Resende City Hall, 2009b.

FIGURE 10- CLINIC BUILDING IN APP

Source: Resende City Hall, 2009c.

There is a certain similarity between the environmental licensing processes in the states of Minas Gerais, Sao Paulo and Rio de Janeiro, in terms of the documentation that must be submitted by the license applicant to the environmental agency. However, none of them make reference in their legislation to the study of impacts on fauna and flora caused by the implementation of housing developments. The issue of geological stability is also ignored in the majority of cases, a fact that causes inconvenience and material losses for the population and the municipality, as can be seen in Figures 11, 12 and 13, when the Sesmaria River flooded in December 2010.

FIGURE 11 - COLLAPSE OF THE WALL OF A LUXURY CONDOMINIUM

Source: Luis Araùjo, 2010.

FIGURE 12 - MIDDLE-CLASS HOUSING CONDEMNED BY CIVIL DEFENSE

Source: Luis Araùjo, 2010.

FIGURE 13 - COLLAPSE OF THE HEAD OF A BRIDGE IN THE CITY CENTER

Source: Luis Araùjo, 2010

With regard to documentation, AMAR requests the same basic documents as for obtaining the Preliminary, Installation and Operating License, namely: completed application form (ANNEX 4), copy of ID card, CPF, CNPJ, Urban Zoning Declaration, property ownership document and floor plan of the development, suppressing, at first, important information such as the descriptive memorial, location plan and plan of the area where the project will be implemented under current conditions and preliminary landscaping plan, earthworks process, among others.

It is during the earthmoving phase that irreversible damage is caused to natural resources. In addition to those mentioned by Freitas et al. (2001), the environmental impacts at this stage include alterations to the natural surface drainage system, the emergence and intensification of erosion and silting, contamination of the soil and surface and underground water, as well as the reduction of flora and fauna habitats. It was during this phase that the main environmental impacts mentioned were observed in the real estate developments studied, B and C.

An analysis of AMAR's environmental licensing process for real estate developments shows that there is a flaw in the application form, as it was drawn up with an industry in mind, due to its high polluting potential. There is therefore no specific form for each activity to be developed, including real estate development, as shown in Annex 4.

Although the process hides item 3 (Planned Interventions), item 4 (Conservation Units), item 5 (Fragile Areas) and item 6 (Water Bodies) provided for in the state process and of great importance for the preservation of biodiversity and natural resources, these items were taken into account by AMAR when applying for the LPs for developments B and C.

Another item also overlooked by most environmental agencies when applying for the LP is the inventory of the fauna and flora of the ecosystem in the region to be subjected to anthropogenic intervention. For Santos (2006), making an inventory of the fauna and flora of a given portion of an ecosystem is the first step towards its conservation and rational use.

Figure 14 shows the land approved for development B. This development, which began in 2008, should have been licensed by the state agency - INEA, due to its characteristics such as the size of the development (52ha) and the fragile areas that exist in the area to be used by this development: wetlands, springs and hilltops, but it was licensed by the Municipality, in accordance with Municipal Decree No. 1.960/2007 (RESENDE, 2007), which provides for Municipal Environmental Licensing. In addition, it is located within a radius, around 8.5km, of a Conservation Unit - PARNA do Itatiaia, in accordance with CONAMA Resolution No. 13/90, in force at the time (BRASIL, 1990b).

FIGURE 14 - DEVELOPMENT AREA B

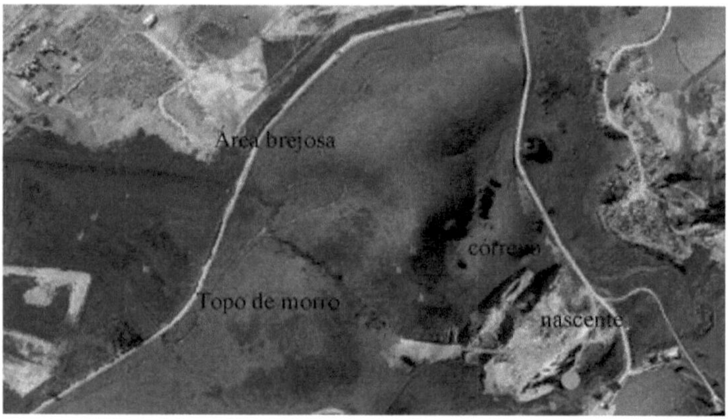

Source: Resende City Hall, 2009d.

The following documents were required for the Municipal Prior License: Zoning Certificate, which indicates whether the site proposed for installation is compatible with the legislation applicable to the use and occupation of municipal land, i.e. basically a zoning certificate in accordance with the Master Plan or the general guidelines for occupation of the territory and the Descriptive Memorial, a document provided by the developer containing a topographical survey marking existing bodies of water, a study of the area's vegetation, occupancy rate, total built-up area, location and land use and occupancy, infrastructure project, preliminary landscaping project and demarcation of the marginal protection strip - FMP, by INEA.

With regard to the documentation requested from the developer, AMAR's concern for the preservation of natural resources can be seen, since it includes respect for the Forest Code, since the request mentions the demarcation of the marginal protection strip, or APP, as well as the topographical survey with the marking of water bodies and a study of the existing vegetation in the area of the development.

Development B, which occupies a hilltop and hillside, has a marshy area and a spring that forms a stream. Despite AMAR's concern for the preservation of natural resources and biodiversity, it became apparent as early as the earthmoving phase in 2010 that the person responsible for the works disregarded the Forest Code and CONAMA Resolution no.303/02, altering the site's water regime by modifying the watershed (hilltop) with earthworks and sealing the soil, which accentuated the environmental impact, damaging the possible flora and fauna that inhabited the site, since freshwater bodies are directly influenced by the processes that occur in the drainage area of the watershed where they are located. We can also see the suppression of the existing spring on the site, influencing the amount of water forming the body of water (FIGURE 15).

FIGURE 15 - ENTERPRISE B

Source: Resende City Hall, 2009e.

Therefore, for project B, it is clear that in the environmental licensing process, although there was concern about the preservation of natural resources due to the requirements of the documentation requested by AMAR, this concern was not fulfilled by the developer due to the damage shown in figure 15, among them: decapitation of the hilltop area, burying of springs and marshy areas.

A similar situation was observed on the island of Santa Catarina (SC) by Reis (2010), where environmental legislation also failed to prevent the occupation of hillsides, highlighting the impotence and omission on the part of the public administration and supervisory bodies in their occupation. As a result, the Atlantic Forest has been separated from the mangroves, preventing the flow of energy and matter between these ecosystems, which is fundamental to their balance.

A similar case, but with a different outcome, was also observed in Campos do Jordao (SP), in the Serra da Mantiqueira, with the construction of a luxury hotel on a hilltop near a spring. In this case, despite the fact that the court decision is still subject to appeal, the hotel was ordered by the courts to be completely demolished with the immediate removal of the rubble and the recovery of the degraded area (MAZZITELLI, 2011).

Project C (figure 16) began its licensing process in 2010 and is being carried out by AMAR, in accordance with the agreement signed with the state. With an area of 42ha, it is currently in the Prior License phase, in which the following documents were required: a plan of the development including water resources, slope of the land, vegetation, direct and indirect influence of the development, topographical survey, location and land use and occupation, demarcation of the Marginal Protection Strip - FMP, by INEA and the Legal Reserve, the latter two for study and release of the LI.

FIGURE 16- DEVELOPMENT AREA C

Source: Resende City Hall, 2009f.

Although AMAR's technical team requested the demarcation of the FMP in the LP, which demonstrates its concern to preserve biodiversity as well as the quantity and quality of the abundant water resources on the site, this fact was ignored by the Resende Environment Agency and the LI was released without the presentation of this document, which shows a lack of commitment to the preservation and conservation of natural resources, including the protection of biodiversity. When INEA noticed environmental degradation during the earthmoving stage, the state asked the municipality in July 2011 to inform the person responsible for the project that the LI had been suspended and that it would be provided by INEA.

Figure 17 shows the suppression of riparian vegetation and figure 18 shows the landfilling of a watercourse and a marshy area, both of which were carried out by the developer at the start of earthworks in March 2011. At the beginning of this study, the project was the subject of several complaints to AMAR regarding environmental degradation, a fact that is proven by figures 18 and 19. Once the environmental damage had been confirmed, the company had its infraction notice annulled by the Agency itself. It is worth noting that no compensation measure was requested for the environmental damage caused, either by the operation of the works or by the damage that was revealed.

FIGURE 17 - SUPPRESSION OF VEGETATION AT THE START OF EARTHWORKS

Source: Luis Araùjo, 2010.

FIGURE 18 - GROUNDING A WATERCOURSE '

Source: Luis Araùjo, 2010.

Suppressing vegetation or preventing it from regenerating directly affects the quality of the water resource, since the soil is removed and is exposed to erosive processes that can lead to siltation of the bodies of water downstream from the development, as well as the quantity of water in rivers and reservoirs and groundwater, as water infiltration into the soil is reduced due to soil sealing and the removal of vegetation, which contributes to the lowering of the water table, affecting not only the basin as a whole, but also the local fauna.

According to Harris et al. (2005), the suppression of the original vegetation makes the soil more susceptible to nutrient leaching, altering its physical characteristics. In addition, the erosive processes that constantly occur on river banks when vegetation cover is removed jeopardize places of shelter and reproduction for species that frequent this habitat, as well as waterfowl that depend directly on forests and the food resources offered by rivers, such as fish and aquatic invertebrates.

According to the author, the rivers silted up by erosion no longer support sufficient populations of prey to continue the food chains involving the region's fauna. The hydrological processes compromised by the removal of vegetation cover, which is largely responsible for the region's biodiversity, will have their flood and drought cycles altered. The bodies of water fed annually by the floods are essential for the survival of species whose preferred habitat is the marshes.

Pigosso et al. (2009) add that the suppression of native vegetation decreases biodiversity by reducing the number of individuals of species that could lose their seed reproduction potential in that environment.

In figure 19, development A shows the use of APP (left side of the stream) to build houses, both on the banks of the stream and around the lagoon. The stability of the soil has also been ignored, flouting the Forestry Code (BRASIL, 1965). The land is being questioned by the State Public Prosecutor's Office and is under trial awaiting the judge's decision on the matter.

FIGURE 19 - DEVELOPMENT A. FMP BUILDING

Source: Resende City Hall, 2009g.

Article 2 of Federal Law No. 4.771/65 stipulates a minimum distance of 30m for bodies of water up to IOm wide. This fact was ignored by the public administration, which allowed homes to be built without the proper distance provided for in the aforementioned law.

Although the State Public Prosecutor's Office has embargoed the plots bordering the stream, the canalization of the stream (Figure 20) by the municipality does not contradict the relevant legislation, i.e. the Forest Code does not define a different length for canalized streams, so the 30m defined by the legislation applies.

An analogous situation was found by Vitte et al. (2010), in the process of urbanization and the incorporation of river banks into the urban space in Campinas SP, which began in the 1930s and are currently responsible for numerous environmental problems.

With the central area densified, the municipality was left to exploit only peripheral allotments. In this sense, the expansion of the urban network is given by the incorporation of areas of native vegetation and permanent preservation, APP, generating predatory environmental practices in the occupation of space.

Among these practices are total soil sealing, the channeling of streams and the suppression of vegetation. According to Mendonça (1994), soil sealing and the canalization of streams are among the most obvious factors in modifying the local climate of cities, as well as violating the Water Code (BRASIL, 1934), which is the oldest restriction on the use of urban land by civil construction and which prohibits any building that could impede or hinder the free flow of water.

Figure 20 shows the canalization of the stream, in May 2011, as a "solution" to the constant flooding of the residential areas built along the left bank of the stream, highlighted in Figure 19.

FIGURE 20 - VIEW OF THE CANALIZATION OF THE STREAM

Source: Luis Araùjo, 2011.

It's important to note that both the channeling of the stream and the interference with the APP (Figure 21) was carried out by the Municipal Works Department without any manifestation from the Environmental Agency and that there were no mitigating or compensatory measures for the environmental damage caused. It is worth pointing out that according to art. 4, § 4 of the Forest Code (BRASIL, 1965) "the competent environmental agency will indicate, prior to issuing the authorization for the suppression of vegetation in a permanent preservation area, the mitigating and compensatory measures that must be adopted by the entrepreneur." In this case, the City Council itself should have adopted the compensatory measures and did not do so.

According to Bechara (2007), the term compensation is used in environmental law to describe different ways of offsetting an environmental loss. The manifesto on environmental compensation is laid down in Article 3 of Federal Law No. 7.347/85, which states that the public civil action may have as its object the fulfillment of an obligation to do or not to do (BRASIL, 1985).

The SNUC Law provides for early environmental compensation, for the purposes of implementing projects that cause significant environmental degradation, with the entrepreneur being obliged to support the implementation and maintenance of a conservation unit from the Integral Protection Group, The amount of resources to be allocated by the entrepreneur for this purpose cannot be less than half a percent of the total costs foreseen for the implementation of the project, according to the degree of environmental impact caused by the project and determined by the licensing body (BRASIL, 2000).

Under the Atlantic Forest Law, the suppression of vegetation in this biome is subject to environmental compensation in the form of the allocation of an area equivalent to the size of the deforested area, with the same ecological characteristics, in the same watershed. However, the law prohibits the suppression of Atlantic forest for the purposes of allotment and building in metropolitan regions and urban areas (BRASIL, 2006).

According to the Forest Code (BRASIL, 1965), the suppression of vegetation in APP can only be authorized in cases of public utility or social interest, duly characterized and motivated in the appropriate administrative procedure, when there is no technical and locational alternative to the proposed undertaking.

Figure 21 shows the suppression of riparian vegetation along the stream (shown in Figure 19) and soil compaction in the APP.

FIGURE 21 - SUPPRESSION OF RIPARIAN FOREST AND SOIL COMPACTION IN APP

Source: Luis Araùjo, 2011.

In practice, Resende's urban growth has not respected the federal legislation protecting the native vegetation of the Atlantic Forest biome, according to Law No. 11.428/2006 (BRASIL, 2006), the Forestry Code (BRASIL, 1965) or its own Municipal Organic Law, which includes among its duties the preservation and restoration of essential ecological processes, the protection of fauna and flora, the prohibition of practices that jeopardize their ecological function and cause the extinction of species, and the reparation of damage caused to the environment by offenders (RESENDE, 1997).

This shows that the municipality disregards the essential value of natural resources, exempting itself from the sovereign right of federal entities over their own resources, as well as denying its responsibilities for conserving them, which goes against the basic principles of the CBD.

Revitalizing streams, rather than channelizing them, is essential if there are to be other uses that are not merely restricted to draining floodwater. With regard to this use, Borsagli & Medeiros (2011)

argue that with channelization, the water starts to flow faster, increasing its flow and taking the problems of flooding to the areas closest to its mouth. Soil sealing, the removal of vegetation and the occupation of floodplains also contribute to aggravating this problem.

On the other hand, according to Melo (2006), streams have a very diverse fauna, made up mainly of various groups of aquatic insects, fish, crustaceans and oligochaetes, which have great potential as bioindicators, ranging from very sensitive species to those that are highly tolerant of pollution, and should therefore be preserved as they are indicators of the state of conservation of the drainage area of the basin to which they belong.

Therefore, the conservation and revitalization of watercourses and their surroundings (APP) in urban areas and the protection of groundwater is also an integral instrument of Water Resources Management that the municipality must promote.

5.4 Conclusion

The environmental licensing process, both carried out by INEA and AMAR in the municipality of Resende (RJ) for real estate developments between 2009 and 2011, did not meet the technical and legal requirements for assessing environmental impacts, thus disregarding the preservation of biodiversity. However, it was noted that the geoprocessing tool called Resende-Geo was fundamental for carrying out this analysis and for verifying and monitoring the state of conservation and preservation of these areas. This has been aggravated by anthropogenic action and the fragmentation of natural *habitats*.

It is understood that some restrictions on land use and occupation should be made by the public body, especially the environmental body responsible for the environmental licensing process, through the following means, among others: mapping and zoning with clear delimitation of the APPs areas, defining the use and occupation of the land in these areas; mapping of the municipality's risk areas, restricting occupation in areas where it is not advisable to build; opposing, on the part of the public body, the release of new developments in areas considered fragile, such as lowlands, hilltops and slopes and setting tax incentives so that the owners of land in APP remain idle or, depending on the case, that they are compensated and that the vegetation in the area is replanted, so that the proposed objectives for the conservation of natural resources established in international legislation and conventions can be guaranteed.

In relation to the environmental licensing process and based on the legal instrument used by the municipality of Resende (RJ) and the findings *in loco*, for the real conservation of natural resources and the quality of life of individuals seeking such enterprises, it is also suggested:

- Application of legislation in all its complexity for the environment to be licensed.

- Effective compliance with the mitigation and compensatory measures presented in the projects.
- According to current legislation, the presentation of recovery plans for degraded areas based on faunal and floristic inventories are essential tools for preserving biotic and abiotic resources.

It is hoped that these measures will strengthen the licensing, inspection and monitoring systems for activities related to biodiversity; create capacity in the body responsible for environmental licensing to assess the impact on biodiversity; support studies into the impact of *habitat* fragmentation on the maintenance of fauna and flora and identify and evaluate public policies that negatively affect biological diversity, thus ratifying the objectives set out in Federal Decree no.4.339/2002, which establishes principles and guidelines for implementing the National Biodiversity Policy.

BIBLIOGRAPHICAL REFERENCES

ABREU, M. A. The city, the mountain and the forest. In: ABREU, M. A. (org). **Nature and Society in Rio de Janeiro**. Rio de Janeiro: Biblioteca carioca , 1992.

AGNES, C. C; CALEGARI, L; GATTO, D. A; STANGERLIN, D. M. **A discussion on the decentralization of environmental management.** Electronic Scientific Journal of Forest Engineering. Year VIII - n. 14. Aug. 2009. Semiannual Periodicals.

ARRUDA, R. **Traditional Populations and the Protection of Natural Resources in Conservation Units.** In Anais do Primeiro Congresso Brasileiro de Unidades de Conservaçao. v. 1. Conferências e Palestras, p. 262-276. Curitiba, Brazil, 1997.

BECHARA, Érika**. A contribution to improving the institute of environmental compensation in Law 9.985/2000**. Thesis (Doctorate in Law). Sao Paulo: Pontifical Catholic University, 2007.

BEGON, M; TOWNSEND, C. R; HARPER, J. L. **Ecology**: From Individuals to Ecosystems. Translated by Adriano Sanches Melo et al. 4.ed. Porto Alegre: Artmed, 2007. Original title: Ecology from Individuals to Ecosystems.

BORSAGLI, A; MEDEIROS, F. G. L. Creeks and the metropolis: The insertion of watercourses that cross the urban area of Belo Horizonte into the urban space. In: **XII Simpurb - National Symposium on Urban Geography**, 2011, Belo Horizonte. The streams and the metropolis: The insertion of the watercourses that cross the urban area of Belo Horizonte into the urban space, 2011.

BESSA, A.. P. **Direito Ambiental**. 8 ed. Rio de Janeiro: Lùmen Juris, 2005.

BRAZIL. Presidency of the Republic. Decree no. 24.643 of July 10, 1934. Decrees the Water Code.

. Law No. 4.771, of September 15, 1965. Institutes the new Forest Code. Brasilia, DF, published in the Official Gazette of the Federative Republic of Brazil on September 15, 1965. Available at: http://www.planalto.gov.br/ccivil/leis/L4771.htm>. Accessed on: September 2, 2010.

. Law n. 6.766, of December 19, 1979. Provides for the subdivision of urban land and other measures. Brasilia, DF, published in the *Official Gazette of the Federative Republic of Brazil* on December 20, 1979. Available at: <http://www.planalto.gov.br/ccivil/leis/L6766.htm>. Accessed on: September 2, 2010.

. Law No. 6.938 of August 31, 1981. Provides for the National Environmental Policy, its purposes and formulation and application mechanisms, and makes other provisions. Brasilia, DF: 1981.

. Presidency of the Republic. Law No. 7.347, of July 24, 1985. Disciplines public civil action for liability for damage caused to the environment, the consumer, goods and rights of artistic, aesthetic, historical, tourist and landscape value (VETOED) and makes other provisions. Brasilia, DF, 1985.

. National Environmental Council. CON AMA Resolution No. 001, of January 23, 1986. Provides for the preparation of the Environmental Impact Study - EIA and respective Environmental Impact Report - RIMA. Brasilia, DF. 1986.

. Presidency of the Republic. Constitution of the Federative Republic of Brazil. Brasilia: 1988.

. Presidency of the Republic. Decree No. 99.274 of July 6, 1990. Regulates Law No. 6.902, of April 27, 1981, and Law No. 6.938, of August 31, 1981, which provide, respectively, for the creation of Ecological Stations and Environmental Protection Areas and for the National Environmental Policy, and makes other provisions.

. National Environment Council. CONAMA Resolution No. 13, of December 6, 1990. Brasilia, DF. 1990.

. National Water Agency. Normative precedents. Brasilia, DF, 1997.

. Presidency of the Republic. Law no. 9.433 of January 8, 1997. Institutes the National Water Resources Policy, creates the National Water Resources Management System, regulates item XIX of art. 21 of the Federal Constitution, and amends art. 1 of Law no. 8.001, of March 13, 1990, which amended Law no. 7.990, of December 28, 1989.

. National Environmental Council. Resolution No. 237, of December 19, 1997. Provides for environmental licensing; competence of the Union, States and Municipalities; list of activities subject to licensing; Environmental Studies, Environmental Impact Study and Environmental Impact Report. DF. 1997.

. Ministry of the Environment. **Convention on Biological Diversity**, Available at: http://www.mma.gov.br/sitio/index.php?ido=conteudo.monta&idEstrutura =72&idConteudo=2870. Accessed on: October 24, 2010.

. Ministry of the Environment. Secretariat for Biodiversity and Forestry. National Register of Conservation Units. Available at: http://www.mma.gov.br/sitio/index.php?ido=conteudo.monta&idEstrutura=119&idConteudo =9677&idMenu=11809. Accessed on: October 24, 2010.

. Presidency of the Republic. Law No. 9.985 of July 18, 2000. Regulates art. 225, § 1, items I, II, III and VII of the Federal Constitution, institutes the National System of Nature Conservation Units and makes other provisions. DF: 2000.

. Presidency of the Republic. Law No. 10.257 of July 10, 2001. Regulates articles 182 and 183 of the Federal Constitution, establishes general guidelines for urban policy and makes other provisions.

. Presidency of the Republic. Provisional Measure No. 2.166-67 of August 24, 2001. Amends arts. 1, 4, 14, 16 and 44, and adds provisions to Law no. 4.771, of September 15, 1965, which establishes the Forest Code, as well as amending art. 10 of Law no. 9.393, of December 19, 1996, which provides for the Tax on Rural Territorial Property - ITR, and makes other provisions.

. National Environment Council. Resolution No. 302, of March 20, 2002. Provides on the parameters, definitions and limits of Permanent Preservation Areas of artificial reservoirs and the regime of use of the surroundings. DF. 2002a.

. National Environment Council. Resolution No. 303, of March 20, 2002. Provides for the parameters, definitions and limits of Permanent Preservation Areas. DF. 2002b.

. Ministry of the Environment. Decree No. 4.297 of July 10, 2002. Regulates art. 9, item II, of Law no. 6.938, of August 31, 1981, establishing criteria for the Ecological-Economic Zoning of Brazil - ZEE, and makes other provisions.

. Presidency of the Republic. Decree No. 4.339 of August 22, 2002. Establishes principles and guidelines for the implementation of the National Biodiversity Policy.

. National Environment Council. Resolution No. 369, of March 28, 2006. Provides for exceptional cases of public utility, social interest or low environmental impact, which allow the intervention or suppression of vegetation in Permanent Preservation Areas (APP). DF. 2006.

. Presidency of the Republic. Law No. 11. 428 of December 22, 2006. Provides for the use and protection of the native vegetation of the Atlantic Forest biome, and makes other provisions. Brasilia, DF: 2006b.

. Ministry of the Environment. Secretariat for Institutional Articulation and Environmental Citizenship. Brasilia. DF. 2007.

. Presidency of the Republic. Decree No. 6.101 of April 26, 2007. Approves the Regimental Structure and the Demonstrative Chart of Commissioned Positions and Gratified Functions of the Ministry of the Environment, and makes other provisions.

. National Water Agency. Available at: http⁄www.2.ana.gov.br. Accessed on: August 20, 2010.

. National Water Agency. **Southeast Atlantic Hydrographic Region**. Available at://www2.ana.gov.br/Paginas/portais/bacias. Accessed on: March 29, 2011.

. Ministry of Cities. Available at http⁄www.cidades.gov.br. Accessed on Aug. 20, 2010.

. Brazilian Institute for the Environment and Renewable Natural Resources. IBAMA Normative Instruction No. 184, of July 17, 2008. Brasilia, DF, 2008.

BRUNER, A; GULLISON, R. E; RICE, R. E; FONSECA, G. A. B. **Effectiveness of parks in protecting tropical biodiversity**. Science, v. 291, p.125-28. 2001.

BUENO, L. M. M. Santo André City Hall. Canadian International Development Agency. **Social Housing in Watershed Areas.** Sao Paulo: Annablume, 2004.

CI-Brasil (Conservaçao Internacional do Brasil). **More protection for biodiversity in southern Bahia.** Available at http://www.conservacao.org/noticias. Accessed on: October 10, 2010.

. **Assessment and Priority Actions for the Conservation of the Biodiversity of the Atlantic Rainforest and Southern Fields**. MMA/SBF, Brasilia. 2000.

COMPANS, R; VINAS, S. M. **The specificities of conflicts involving irregular occupations in areas considered unsuitable for urbanization.** Rio de Janeiro, n. 16-17, May/Dec. 2005.

COSTA, H. S. M; PEIXOTO, M. C. D. **Real estate dynamics and environmental regulation:** a discussion based on the southern axis of the Belo Horizonte Metropolitan Region. v.24, n.2, p. 317336, jul./dez. 2007.

CRUZ, R. C. A. **Introduçao à Geografia do Turismo**. Sao Paulo: Roca, 2001.

CURICACA, **Judge upholds lawsuit against power plant in Cidreira.** Available at: http://ong.portoweb.com.br/curicaca. Accessed on March 25, 2010.

FARIA, H. H. Evaluating the Effectiveness of Conservation Unit Management: how to proceed? In: **Brazilian Congress of Protected Areas**. Curitiba, RNPUC, SEMA, FBPN, Anais v. 1, p. 478-499, 1997.

FARIAS, T. Q. Aspectos gerais da politica nacional do meio ambiente - comentârios sobre a Lei n° 6.938/81. In: **Àmbito Juridico**, Rio Grande, 2006. Available at: http://www.ambito-juridico.com.br/revistas_artigos_leitura. Accessed on: March 1, 2011.

FENKER, E. A. Cost-benefit analysis applicable to the environment. Available at: http://noticias.ambientebrasil.com.br/artigos/2007/01/22/28939-valor-economico-da-biodiversidade.html. Accessed on: 02/09/2010.

FERRAZ, K. **Forest Reserves call for help**. Conhecimento pràtico geografia: Sao Paulo: n. 30, p. 52-61, April 2010.

FRANK, B. et al. **A Reflection on Tragedy in Santa Catarina**. 2008. Available at: http://www.apremavi.org.br/noticias/apremavi. Accessed on: November 24, 2010.

FREITAS, C. G. L. (coordinator); BRAGA, T. O; BITAR, O. Y; FARAH, F. **Habitaçao e meio ambiente** - Abordagem integrada em empreendimentos de interesse social. Sao Paulo: Technological

Research Institute of the State of Sao Paulo - IPT, 2001.

GUSMÂO, A. C. **Environmental Legislation, Management and Responsibilities**. Handout for the subject Environmental Law. LATEC-UFF, 2008. Brazilian Institute of Geography and Statistics - IBGE. Statistical Yearbook. Brasilia: DF, 2004.

HARRIS, M. B.; ARCANGELO, C.; PINTO, E.C.T.; et. al. **Estimates of natural area loss in the Upper Paraguay river basin and the Brazilian Pantanal**. Unpublished technical report. Conservaçao Internacional, Campo Grande: MS, 2005

HARAGUCHI, L. M. M; CARVALHO, O. B. (Org.). **Medicinal Plants**. Municipal Department of Greenery and the Environment. Technical Division Municipal Gardening School. Sao Paulo: 2010.

HOLANDA, S. B. **Raizes do Brasil**. 26 ed. Sao Paulo: Cia das Letras, 1995.

BIOMES INSTITUTE. **Strategies and actions for the conservation of Biodiversity in the State of Rio de Janeiro**. Rio de Janeiro: 2009.

BRAZILIAN INSTITUTE OF GEOGRAPHY AND STATISTICS, Ministry of Planning, Budget and Management. 2010 Census.

. Available at: www.ibge.gov.br/cidadesat. Accessed on: September 19, 2010.

. IBGE launches the Map of Biomes of Brazil and the Map of Vegetation of Brazil, in commemoration of World Biodiversity Day. Available at: http://www.ibge.gov.br/home/presidencia/noticias/noticia. Accessed on: September 19, 2010.

BRAZILIAN INSTITUTE OF THE ENVIRONMENT AND RENEWABLE NATURAL RESOURCES. **Protect and Conserve**: Everyone's Obligation. Available at: http://www.ibama.gov.br/fauna/home.htm. Accessed on: October 24, 2010.

. Normative Instruction no. 184 of July 17, 2008.

STATE ENVIRONMENTAL INSTITUTE. Decentralization of Environmental Licensing in the State of Rio de Janeiro. Rio de Janeiro. INEA, 2010.

ETHOS INSTITUTE. **Companies launch movement for the conservation and sustainable use of biodiversity.** Available at: http://www1.ethos.org.br/noticias. Accessed on: February 22, 2011.

INTERNATIONAL ENVIRONMENTAL TECHNOLOGY CENTER. **Planning and Management of Lakes and Reservoirs**: an integrated approach to the problem of eutrophication. UNEP-IETC. Original title: Planning and Management of Lakes and Reservoirs: An Integrated Approach to Eutrophication, 2001.

RIO DE JANEIRO BOTANICAL GARDEN RESEARCH INSTITUTE. Available at:

http://www.jbrj.gov.br/pesquisa/div_tax/briofitas/mapas.htm. Accessed on: December 12, 2010.

JURISWAY EDUCATIONAL System. **What are allotments**? [S. l.], 2009. Available at: <http://www.jurisway.org.br/v2/pergunta.asp?idmodelo=8911>. Accessed on: September 1, 2010.

LIMA e SILVA, P. P. GUERRA, A. J. T. MOUSINHO, P. BUENO, C. ALMEIDA, F. G. MALHEIROS, T. M. M. SOUZA JUNIOR, A. B. **Dicionârio Brasileiro de Ciências Ambientais**. Thex. 2.ed. Rio de Janeiro. 2002.

LIMA, W. P. **Forest hydrology applied to watershed management**. Piracicaba - SP: University of Sao Carlos, ESALQ, Department of Social Sciences, 1996.

MAGALHÂES, C. S; FERREIRA, R. M. A. **Permanent preservation areas in a micro-basin.** Informe Agropecuârio. Belo Horizonte, v.21, n.207, p.33-39, nov/dec. 2000.

MARTINS, S. V. **Recuperaçao de Matas Ciliares**. 2.ed. Viçosa, MG, 2007.

MAZZITELLI, F. **Justice orders demolition of luxury hotel in environmental area**. The State of Sao Paulo. Sao Paulo. June 19, 2011. Available at: http://www.estadao.com.br/noticias/cidades. Accessed on: June 29, 2011.

MEDAUAR, O. (Org.) **Coletânea de Legislaçao Ambiental**. Revista dos Tribunais. 10ª ed. Sao Paulo: 2011.

MEDEIROS, J. D. Avaliaçao de Impacto Ambiental In: **Reuniao anual da sociedade brasileira para o progresso da ciência**, 47, 1995, Anais.

MELO, A. S. Diversity of macroinvertebrates in streams. In: **Método de estudos em Biologia da Conservaçao Manejo da Vida Silvestre**. CULLEN JUNIOR, L; RUDRAN, R; VALLADARES-PADUA, C. (Orgs). 2 ed. Curitiba: Universidade Federal Paranà, 2006.

MENDONÇA, F. A. **O Clima e o Planejamento Urbano das Cidades de Porte Médio e Pequeno**: Proposições Metodológicas para Estudo e sua Aplicaçao à Cidade de Londrina/PR. 1994. Sao Paulo: Thesis (Doctorate) Postgraduate Program in Physical Geography, Faculty of Philosophy, Letters and Human Sciences, University of Sao Paulo, Sao Paulo: 1994.

MILARÉ, E. **Direito do Ambiente**: doutrina - pràtica - jurisprudência - glossàrio. 2.ed. Sao Paulo: Revista dos Tribunais, 2004.

MINAS GERAIS. **COPAM Normative Decision No. 58, of November 28, 2002.** Establishes rules for the environmental licensing of urban land subdivisions for exclusively or predominantly residential purposes, and makes other provisions.

MINAYO, M. C. S. MACHADO, J. M. H. MATOS, L. B. F. ODA, L. M. VIEIRA, V. M.

MONTEIRO, T. C. N. **Fiocruz Saudâvel:** uma experiência institucional. Ciência e Saùde Coletiva, Rio de Janeiro, v.3, n.2, p. 151-161, 1998.

MORIN, E. **Os sete saberes necessârios à educaçao do futuro**. traduçao de Catarina Eleonora F. da Silva e Jeanne Sawaya. - 2. ed. Sao Paulo: Cortez, Brasilia, DF: UNESCO, 2000. Original title: Les sept savoirs nécessaires à l'éducation du futur.

PIGOSSO, M; BONFANTE, E. FARIAS, E; BECEGATO, V; ONOFRE, S. B. **Diagnóstico Ambiental da Bacia Hidrogràfica do Rio Jirau**. Geoambiente *on line*. Jatai. n° 13. jul-dez. 2009.

PINHEIRO, L.A. F. V; VIANA, V. M. **Conservaçao da biodiversidade em fragmentos florestais.** IPEF. v. 12, n. 32, p. 25-42, dec. 1998.

PINTO, L. P; BEDÊ, L.; PAESE, A. FONSECA, M. PAGLIA, A. LAMAS, I. **Brazilian Atlantic Forest**: The Challenges of Preserving the Biodiversity of a Global *Hotspot*. Available at: http://dc145.4shared.com/doc/9zmJLzBt/preview.html. Accessed on: October 10, 2010.

PINTO, V. C. **Ocupaçâo Irregular do Solo Urbano**: o papel da legislaçao federal. p. 1-13, 08 Available at: http://www.senado.gov.br/senado/conleg/artigos/direito/OcupacaoIrregulardoSoloUrbano.pdf. Accessed on: October 10, 2010.

PORTO ALEGRE. Complementary Law No. 434 of December 1, 1999. Provides for urban development in the Municipality of Porto Alegre, institutes the Porto Alegre Urban Environmental Development Master Plan and makes other provisions. Porto Alegre, 1999.

PRIMACK, Richard B. RODRIGUES, Efraim. **Conservation Biology**. Londrina: Planta, 2006.

RICKLEFS, R. E. **The Economy of Nature**. Translated by Cecilia Bueno and Pedro Paulo de Lima e Silva. Rio de Janeiro. Guanabara Koogan. 6.ed. 2010. Original title: The Economy of Nature: A Texbook in Basic Ecology.

RESENDE. Resende Municipal Council. Organic Law of the Municipality. 1997. ______. Resende City Hall. Map of the Municipality of Resende. 2010. Available at: http://www.resende.rj.gov.br/page/mapa.asp. Accessed on: October 1, 2010. ______. Resende City Hall. Municipal Decree No. 1.960 of November 7, 2007. Provides for Municipal Environmental Licensing. Resende. 2007. ______. Resende City Hall. Geographic Information System for Territorial Management. 2009a.

______. Resende City Hall. Geographic Information System for Territorial Management. 2009b.

______. Resende City Hall. Geographic Information System for Territorial Management. 2009c.

______. Resende City Hall. Geographic Information System for Territorial Management. 2009d.

______. Resende City Hall. Geographic Information System for Territorial Management. 2009e.

______. Resende City Hall. Geographic Information System for Territorial Management. 2009f.

______. Resende City Hall. Geographic Information System for Territorial Management. 2009g.

REIS, A. F. **Environmental Preservation in the Urban Context**: City and Nature on the Island of Santa Catarina. R. B. Estudos Urbanos e Regionais. v. 1 2, n .1. m a y 2 0 1 0.

UNITED NATIONS. **Rio Declaration on Environment and Development**. Rio de Janeiro:

1992.

RIO DE JANEIRO. Decree-Law no. 134 of June 16, 1975. Provides for the prevention and control of environmental pollution in the state of Rio de Janeiro and other measures.

______. Decree-Law no. 1.633 of December 21, 1977. Partly regulates Decree-Law No. 134, of 16-06-75, and establishes the Polluting Activities Licensing System.

______. Law no. 1.356 of October 3, 1988. Provides for procedures related to the preparation, analysis and approval of environmental impact studies. Rio de Janeiro, 1988. ______. Law No. 5.067 of July 9, 2007. Provides for the ecological-economic zoning of the state of rio de janeiro and defines criteria for the implementation of economic forestry activities in the state of rio de janeiro. Rio de Janeiro, 2007. ______. Decree No. 42.050 of September 25, 2009. Regulates the procedure for decentralizing environmental licensing by signing agreements with municipalities in the state of Rio de Janeiro, and makes other provisions. Rio de Janeiro, 2009. ______. Decree No. 42.356 of March 16, 2010. Provides for the treatment and demarcation of marginal protection strips in environmental licensing processes. Rio de Janeiro, 2010.

RIO GRANDE DO NORTE. Institute for Economic Development and the Environment of Rio Grande do Norte. **Ecological-Economic Zoning Project for the Estuaries of Rio Grande do Norte - ZEE/RN.** Natal. 2005.

RODRIGUES JUNIOR, J. B. Programa Social Berimbau de auto-sustentabilidade da Costa do Sauipe: **Avaliação do Programa na Comunidade de Porto do Sauipe**. Brasilia: Faculty of Economics, Administration, Accounting and Science of Information and Documentation. 2006. 125p. (Dissertation, Master's Degree in Social Management).

RYLANDS, A. B; BRANDON, K. **Brazilian conservation units**. Belo Horizonte: v. 01, n. 01, p. 28-35, July 2005.

SANTOS, A. A. S; GÓES, J. A. **Considerations on the preservation of biodiversity and the role of Law No. 11.284/06.** Jus Navigandi, Teresina, year 11, n.1036, May 2006. Available at: http://jus.uol.com.br/revista/texto/8341. Accessed on: Feb. 26, 2011.

SANTOS, A. J. Estimates of species richness. In: **Método de Estudos em Biologia da Conservaçâo Manejo da Vida Silvestre**. CULLEN JUNIOR, L; RUDRAN, R; VALLADARES-PADUA, C. (Orgs). 2 ed. Curitiba: Universidade Federal Paranà, 2006.

SANTOS, R. F. **Environmental Planning**: Theory and Practice. Oficina de Textos. Sao Paulo: 2004.

SANTOS FILHO, P. S. Habitat fragmentation: implications for in situ conservation. In: ESTEVES, F. D. A. (ed.) **Structure, Functioning and Management of Brazilian Ecosystems**. Rio de Janeiro: PPGE-UFRJ, pp. 365-393, vol. 1, 1995.

São Paulo State Department of the Environment, Forestry Foundation. **Nature Conservation Units.** Organizer: Luiz Roberto Numa de Oliveira. Sao Paulo: SMA, 2009. 104 p.

SÂO PAULO. Decree No. 49.215, of December 7, 2004. Provides for the Ecological-Economic Zoning of the North Coast Sector, foresees uses and activities for the different zones, establishes guidelines, environmental and socio-economic goals, under the terms established by Law No. 10.019, of July 3, 1998.

São Paulo State Department of the Environment. Botanical Institute; coordinator: Vera Lùcia Ramos Bononi. **Biodiversity.** Sao Paulo: SMA, 2010.112p.

SERRA, C. A. T. **Geography, Ecology and the Environment**. Collection of Articles. 2.ed. Rio de Janeiro: Princeps Digital, 2003.

SERVILHA, É. R. et al. **Permanent preservation areas, cities and the urban**.

Environmental Law Magazine, Sao Paulo, 2005. Available at: http://www.e-science.unicamp.br/fluxus/admin/publicacoes/documentos/publicacao_590_Revista_DIreito_%20Ambiental%20.pdf. Accessed on: January 13, 2010.

MUNICIPAL ENVIRONMENTAL SANITATION SERVICE. **Guidelines for drawing up an environmental control plan (PCA).** Sao Paulo.

SOS MATA ATLANTICA. Atlantic Forest. **Atlantic Forest**. Available at: http://www.sosmatatlantica.org.br/index.php?section=info&action=mata. Accessed on: October 10, 2010.

. **Atlas Remanescentes** 2008. Available at: http://mapas.sosma.org.br/. Accessed on: October 10, 2010.

SOUSA, D. V. **The Concept of Environmental Impact within the Framework of the Concept of Sustainability.** Caminhos de Geografia. v. 6, n. 19. p. 126-129 Page 126. Uberlândia: 2006.

SOUZA, J. A; CUNHA, M. A. et al. **Disordered Occupation**. Conhecimento Pràtico Geografia. Sao

Paulo: n. 30, p.32-35, April 2010.

SOUSA, J. P. **Biodiversity and the Economy**: Some Reflections. Available at: http://www.ecoeco.org.br/conteudo/publicacoes/encontros/iii_en/Mesa8/5.pdf. Accessed on: October 10, 2010.

STEINBERGER, M. **Zoneamento Ecolôgico-Econômico como instrumento de ordenamento territorial urbano e regional**. Revista Agora, Brasilia: Paralelo 15, v. 1, n1, 1997.

STROHAECKER, T. M. **Population Dynamics**. Macro-diagnosis of the Brazilian Coastal and Marine Zone: IBAMA/MMA, p.59-73, 2008.

TARGA, M. S. **Impactos Ambientais em Bacias Hidrogràficas**. Taubaté, Postgraduate Program in Environmental Sciences, University of Taubaté, Taubaté, 2008. 61p.

UEZU, A. Use of geographic information systems in conservation biology. In: **Mètodo de estudos em Biologia da Conservaçao Manejo da Vida Silvestre**. CULLEN JUNIOR, L; RUDRAN, R; VALLADARES-PADUA, C. (Orgs). 2 ed. Curitiba: Universidade Federal Paranà, 2006.

UNESCO. **CONVENTION FOR THE PROTECTION OF WORLD CULTURAL AND NATURAL HERITAGE**. Available at: http://whc.unesco.org/archive/convention- en.pdf. Accessed on: September 15, 2010.

Economic Value of Biodiversity. Available at

http://noticias.ambientebrasil.com.br/artigos/2007/01/22/28939-valor-economico-da-biodiversidade.html. Accessed on: September 20, 2010.

VILLARES, L. F. **CONAMA's normative power**. Rev. Juridica. Brasilia, v. 10, n. 90, Ed. Esp., p.01-11, Apr./May, 2008.

VITTE, A. C. CISOTTO, MARIANA F. VILELA FILHO, L. R. **A urbanizaçao e a incorporaçao das vàrzeas ao espaço urbano de campinas** (SP), BRASIL. Revista Geografar. Curitiba: v.5, n.1, p.105-132, jan./jun. 2010.

VIVA FAVELA. **Number of slum dwellers in Brazil will be 55 million in 2020.** Available at: http://www.vivafavela.com.br/materias. Accessed on: October 20, 2011.

VERDADE, V. K; DIXO, M; CURCIO, F. F. **The risks of extinction of frogs, toads and tree frogs as a result of environmental changes.** Estudos Avançados. v.24. n. 68, Sao Paulo: 2010.

WALCACER, F. A nova lei de loteamentos. In: PESSOA, A. (Coord.). **Direito urbanistico**: uma viso socio-juridica. Rio de Janeiro: Livros Técnicos e Cientificos / IBAM, 1981.

ANNEXES

Annex 1 - Environmental licensing in the state of Rio de Janeiro for real estate developments:

Among the projects subject to environmental licensing are buildings, allotments and land development projects, urban drainage works and road paving, cuts and embankments for leveling grids.

6.38 REAL ESTATE, TOURIST AND LEISURE DEVELOPMENTS AND AREAS FOR EXCLUSIVELY INDUSTRIAL USE

This includes residential and commercial buildings, residential or industrial subdivisions, housing estates, tourist complexes, theme parks, strictly industrial zones and industrial districts.

Size	Area of intervention (m2)
Minimum	Up to 2,000
Small	Above 2,000 to 20,000
Medium	Over 20,000 to 100,000
Large	Over 100,000 to 500,000
exceptional	Above 500,000

Weight	Conditioning factor	Situation	value
10	Located in or around a fragile area (Annex 1)	no yes	0 1
10	Provides for cuts and embankments	no yes	0 1
10	Provides for alterations to bodies of water or modifies natural drainage	no yes	0 1
8	Provides for the removal of vegetation	no yes	0 1

Polluting potential	Sum of weight x value parcels
Bass	0 a 18
Medium	19 a 35
High	36 a 43

Annex 1

FRAGILE AREAS

- Slopes or parts thereof with a gradient equal to or greater than 24% (twenty-five percent).

- Slopes with a gradient equal to or greater than 10% (ten percent) in coastal areas.

- Woodlands or Forests - complex ecosystems in which trees are the predominant plant form that protect the soil from the direct impact of the sun, wind and precipitation.

- Restingas - sandy coastal accumulations, parallel to the coastline, generally elongated in shape, produced by sediments transported by the sea, where characteristic mixed plant associations are found, commonly known as "restinga vegetation".

- Dunes - sandy coastal accumulations produced by the action of the wind in whole or in part, stabilized or fixed by vegetation.

- Wetlands - land that is wet or saturated with water, sometimes flooded from time to time, covered with its own natural vegetation in which shrubs integrated with creeping grasses and some tree species predominate.

- Mangroves - "coastal ecosystems" that occur on low-lying land subject to the action of the tides, located in relatively sheltered areas such as bays, estuaries and lagoons and are usually made up of recent mud flats, which are associated with particular types of flora and fauna.

- Areas of endemism - isolation of one or many species in a terrestrial space, following genetic evolution that differs from that which occurred in other regions.

- Areas that are home to endangered species.

- Archaeological sites - areas designed to protect traces of prehistoric human occupation from any alterations and where activities are disciplined and controlled so as not to harm the values to be preserved.

- Areas of influence of springs or waterholes, reservoirs, river courses, lakes, lagoons and beaches.

1- BUILDINGS

Preliminary License - LP

Simplified Environmental Registration - Various projects

Descriptive memorial containing:

- criteria that guided the party adopted to justify the relocation of the contour lines;

- occupancy rate (O.R.);

- Total built-up area (TAE)

- Project population and estimated population densities (net and gross);

- preliminary sizing of the areas intended for the different uses (housing, recreation and leisure, parking, commerce and services, social and sporting activities, security and others). This item does not apply to multi-family residential developments;

- indication of the stages envisaged in the case of modular deployment.

Projected road layout

Location plan of the development, on a scale compatible with the table of minimum scales below, indicating the following elements:

- magnetic orientation;

- projected topography;

- existing and planned bodies of water and their respective buffer strips;

- vegetation cover, including that considered permanent preservation by the Forest Code;

- road system and planned accesses;

- projection of buildings and areas intended for the different uses envisaged

development area	minimum scale
up to 2,000 m^2	1:250
between 2,000 and 10,000 m^2	1:500
between 10,000 and 50,000 m^2	1:1.000
over 50,000 m^2	1:2.000

Plans of the buildings on a scale of at least 1:500, with a graphic indication of the spatial distribution of the equipment and the covered and uncovered areas intended for the different uses.

Cross-sections and longitudinal sections showing the development on the site.

Information on sanitation infrastructure, including:

- water supply system;

- sewage system;

- rainwater drainage system;
- solid waste collection and disposal.

If there is a need to suppress native vegetation or intervene in a permanent preservation area as classified by Federal Law No. 4.771/65 of 15/09/65 and CONAMA Resolution No. 303 of 20/03/02, click here to view the list of documents to be submitted.

Installation License - LI

Simplified Environmental Registration - Various projects

Declaration from the sewage utility about the possibility of connecting to the network.

Road system design.

Sanitation infrastructure projects:

- water supply system;
- sewage system;
- rainwater drainage system;
- solid waste collection and disposal.

Declaration from the competent body on the possibility of collecting solid waste.

Landscape design.

Documents listed in the LP to be submitted with the LI application.

In the case of the use of state-owned water resources, present proof of the application or the document granting the right to use water resources, or a declaration of insignificant use. If there is no application or grant, click here to view the procedures to be followed to obtain the grant.

If there was no LP:

Descriptive memorial and plans required for granting the LP.

Related Standards

- IT-1814.R-5 - Technical Instruction for the Presentation of Building Plans

Multifamily residences (permanent or temporary), group buildings and clubs.

- IT-1815.R-5 - Technical Instruction for the Presentation of Residential Building Projects

Multifamily (Permanent and Transitional), Groups of Buildings and Clubs, approved by CECA Resolution No. 868 of 08.05.86 and published in the D.O.R.J. of 19.05.86.

These standards can be found on this website, under **Environmental Legislation and Standards**.

Annex 2 - Simplified Environmental Registration and documentation

inea instituto estadual do ambiente	Environmental Licensing System - SLAM SIMPLIFIED ENVIRONMENTAL REGISTRATION MISCELLANEOUS WORKS	PROCEDURE: FL: RUBRIC

1- Identifying the Business/Activity

Name/Business name

II

CNPJ

Activity location

Neighborhood Municipality ZIP CODE

2- Characteristics of the Location of the Activity or Enterprise

Total area:m² Intervention area:m²

Coordinates

Geographical - Longitude: Latitude:

UTM - XY

Zoning:

3- Planned interventions

( ) Tree felling ( ) Intervention in

Vegetation IsolatedAPP

( ) Cut and fill ( ) Other: Specify

4- Conservation Units

Is the activity or enterprise located inside or within 10 km of Conservation Units?

( ) No ( ) Yes Name and distance from the unit:

5- Fragile Areas

The Activity or Development is located in fragile areas (defined in the Annex to MN0050)

( ) No ( ) Yes Type:

6- Water bodies

Are there any water bodies in or around the intervention area?

( ) No ( ) Yes Name of watercourse

7- Water consumption

Quantity: m /day³

( ) Public network ( ) Groundwater well ( ) Deep well ( ) Surface spring

( ) Other, please specify ______________________________

8- Sewage Treatment

Number of people assisted:

Type of treatment

( ) Pit ( ) Pit and Biological Filter ( ) Secondary Treatment ( ) Treatment in third-party systems ( ) Other, please specify

9- Declaration

I declare, under penalty of law, that the information contained herein is true.

Rio de Janeiro, from

Signature of legal representative

Name of legal representative

Documentation required to license allotments:

Prior License - LP

Simplified Environmental Registration - Various projects

1- Descriptive memorial containing:

- criteria that guide the adopted layout, with justification for the relocation of the contour lines;
- occupancy rate (O.R.);
- Total built-up area (TAE);
- Project population and estimated population densities (net and gross);
- preliminary dimensioning of the areas destined for the different uses envisaged (housing, recreation and leisure, parking, commerce and services, social and sporting activities, security and others);
- indication of the stages envisaged in the case of modular deployment;

- projected road layout.

2- Location plan of the area to be plotted, on a scale commensurate with the size of the development, at least 1:10,000, graphically indicating the following elements within 500 meters:

- magnetic orientation;
- topography;
- bodies of water;
- vegetation cover;
- areas specially protected by legislation;
- uses implemented;
- accesses.

3- Plan of the area where the project will be implemented under current conditions, on a scale commensurate with the size of the development, at least 1:2,000, graphically indicating the following elements:

- magnetic orientation;
- topography, highlighting contour lines every 5 meters;
- existing and planned bodies of water and their respective buffer strips;
- vegetation cover, including that considered permanent preservation by the Forest Code;
- existing roads;
- existing buildings;
- indication of the areas for the various uses envisaged.

4- Preliminary plot plan on a scale compatible with the size of the development, at least 1:1,000, graphically indicating the following elements: - magnetic orientation;

- projected topography with relocated contour lines;
- location of green areas, preservation areas, recreation areas, archaeological sites, historical monuments and others; - protection strip for bodies of water;
- location of blocks and plots clarifying: types and rate of occupation, density, common use buildings and residential units planned as an integral part of the development;
- road system.

5- Information on sanitation infrastructure, including:

- water supply system;
- sewage system;
- rainwater drainage system;
- solid waste collection and disposal.

6- Landscape project.

If there is a need to suppress native vegetation or intervene in a permanent preservation area, as classified by Federal Law No. 4.771/65 of 15/09/65 and CONAMA Resolution No. 303 of 20/03/02, **click here** to view the list of documents to be submitted.

Installation License - LI

Simplified Environmental Registration - Various projects

Declaration from the sewage utility about the possibility of connecting to the network.

Road system design.

Sanitation infrastructure projects:

- water supply system;
- sewage system;
- rainwater drainage system;
- solid waste collection and disposal.

Environmental protection measures:

- in terms of hillside erosion;
- in works carried out on rivers and canals;
- in the opening of channels;
- siltation and undermining of beaches.

Landscaping project. Documents listed in the LP to be submitted with the LI application.

In the case of the use of state-owned water resources, present proof of the application or the document granting the right to use water resources, or a declaration of insignificant use. If there is no application or grant, **click here** to view the procedures to be followed to obtain the grant.

If there was no LP:

Descriptive memorial and plans required for granting the LP.

Related Standards:

- IT-1818.R-4 - Technical Instruction for the Presentation of Preliminary Land Development Projects.

- IT-1819.R-4 - Technical Instruction for the Presentation of Land Development Projects.

Annex 3 - Guidelines for applying for an Environmental License in Resende, RJ Guidelines for applying for an Environmental License

Environmental licensing is the administrative procedure by which the Resende Environment Agency - AMAR authorizes the location, installation, expansion and operation of undertakings and activities that use environmental resources and are considered to be effectively or potentially polluting and those capable, in any form, of causing environmental degradation.

In the analysis of the environmental licensing process, the impacts caused by the project are assessed, such as its potential or capacity to generate liquid effluents, solid waste, emissions into the atmosphere and noise, as well as its risk potential, such as the possibility of causing explosions and fires.

There are three stages in environmental licensing, each of which corresponds to a type of license:

PRIOR LICENSE - LP

Granted in the preliminary planning phase of the project, it authorizes its location, based on federal, state and municipal land use plans, and establishes the basic requirements to be met in the implementation and operation phases.

INSTALLATION LICENSE - LI

Authorizes the start of implementation of the project, in accordance with the specifications of the engineering project and specifies the environmental requirements to be followed during this phase. Authorizes the start of project implementation

OPERATING LICENSE - LO

Issued after verification of compliance with the conditions of the Installation License (LMI), it authorizes the operation of the activity, provided that the specified conditions are respected.

In all three phases of licensing, the license granted by AMAR must be renewed 120 days before its expiry date (License Renewal).

Documents required to apply for an Environmental License

1. Fill in the application form and attach the general documents;

Environmental License Application - **click here**

GENERAL DOCUMENTS

- Application form completed and signed by the legal representative;

• Copies of the ID and CPF of the legal representative signing the application. If the applicant is an individual, they must also provide proof of residence.

• If you have an attorney-in-fact, present a copy of the power of attorney, public or private, with a notarized signature, and a copy of your identity document and CPF number.

• Copy of the minutes of incorporation and election of the last board of directors, in the case of a S/A, or articles of association in the case of a limited liability company. If the applicant is a public body, the Act appointing the representative signing the application must be submitted.

• Copy of registration in the National Register of Legal Entities (CNPJ).

• Copy of the result of the technical consultation prior to the urban indices location permit, which will state that the location and type of enterprise or activity are in conformity with the legislation applicable to land use and occupation.

• Location plan, in copy of IBGE plans, aerial photo, satellite image or sketch, indicating:

• Geographical coordinates

• Location of the land in relation to the main road and at least two others, indicating the name of the accesses; if it is located on the side of a road or highway, indicate the kilometer and the side on which it is located;

• Location of the bodies of water (rivers, springs, ponds, lakes, etc.) closest to the development, with their respective names, if any;

• Uses of neighboring properties and areas, within a radius of at least 100 meters.

2. Schedule a meeting with the person in charge of AMAR's environmental licensing department by calling (024) 3354 7792/8663;

3. At the appointed time, attend to clarify any doubts about the general documentation submitted, and according to the characteristics of the enterprise or activity, receive a list of the specific documentation required;

4. After the general documentation has been analyzed, a technical inspection will be scheduled on site within a maximum of 30 days.

5. Other information and specific documents may be required on a case-by-case basis.

Annex 4 - Application for environmental licensing

amar
Agência do Meio Ambiente do Município de Resende

APPLICATION FOR ENVIRONMENTAL LICENSE

SMALL AND MEDIUM-SIZED

1- IDENTIFICATION

SOCIAL REASON:	
NAME OF LEGAL GUARDIAN:	
CNPJ:	CPF:
ADDRESS:	N°: ZIP CODE:
COMPLEMENT:	BAIRRO:
MUNICIPALITY:	PHONE: ( ) -
E-MAIL:	FAX: ( ) -

2- REQUEST

( ) PRELIMINARY LICENSE - LP	( ) INSTALLATION LICENSE - LI
( ) OPERATING LICENSE - LO ( ) OTHER:	
() RENEWAL: NO PREVIOUS LEAVE: ______________________________	

3- DESCRIPTION OF THE ACTIVITY

4- AREA ALLOCATED TO THE ACTIVITY (m)^2

LAND:	BUILT:
OUTDOOR ACTIVITY:	EQUIPMENT:
ADDRESS:	

LATITUDE:	LONGITUDE:	ALTITUDE:

5- GEOGRAPHICAL COORDINATES

6- PRODUCTION PERIOD AND NUMBER OF EMPLOYEES

OPENING hours: START	END
N° TOTAL NUMBER OF EMPLOYEES	

ADMINISTRATIVE SECTOR:	PRODUCTION SECTOR:

7- USE OF ENVIRONMENTAL RESOURCES

- WATER USE

( ) PUBLIC SUPPLY

(X) OTHER:

- LAND USE AND OCCUPATION - ZONING

( ) INDUSTRIAL ZONE - ZI

( ) SPECIAL ZONE - ZE

(X) OTHER:

8- LIST OF RAW MATERIALS

DESCRIPTION	AVERAGE ANNUAL QUANTITY (UNIT OF MEASUREMENT)

9- LIST OF PRODUCTS

DESCRIPTION	AVERAGE QUANTITY ANNUAL(UNIT OF MEASUREMENT)

10- INVENTORY OF WASTE GENERATED (Liquid, Solid and Air Pollutants)

WASTE DESCRIPTION	PLACE OF DISPOSAL	AVERAGE MONTHLY QUANTITY

11- I DECLARE THAT THE PERFORMANCE OF THE ACTIVITIES RELATED IN THIS APPLICATION WILL BE ACCORDING TO THE DATA PROVIDED, AND I **HEREBY** REQUEST THE ISSUANCE OF THE MUNICIPAL ENVIRONMENTAL LICENSE.

REsENDE, ____________________ DE ____________________________ DE ________ .

LEGAL REPRESENTATIVE'S SIGNATURE

MIX
Papier aus verantwortungsvollen Quellen
Paper from responsible sources
FSC® C105338

Printed by Books on Demand GmbH, Norderstedt / Germany